Bees, wasps and ants

of Kent

A provisional atlas by

G. W. Allen

Published by the Kent Field Club

Published by Kent Field Club, the Natural History Society of Kent

The aims of the Kent Field Club are to promote an increased interest in natural history, to study and record the distribution and ecology of fauna and flora, and to promote nature conservation in association with the relevant organisations within the County of Kent.

First published 2009

© G W Allen

All rights reserved.

Except for copying of small parts for private study or review (as permitted under the copyrights acts), no part of this publication may be reproduced, stored in a retrieval system, or transmitted in any form or by any means, electronic or otherwise, without the prior permission of the publishers and copyright holders.

© Natural England material is reproduced with the permission of Natural England

Maps based on Ordnance Survey material. © Crown Copyright. Reproduced by permission of Ordnance Survey®.

 Kent Field Club wishes to acknowledge the valuable assistance given by the Kent and Medway Biological Records Centre in the production of this book.

ISBN 978-0-9561926-0-8

Cover photograph: *Bombus lucorum* © R I Moyse

CONTENTS

Summary 1
Introduction 1
    Acknowledgements 1
    Recording area 2
    Classification of the organisms 2
    A Resumé of Life Histories 2
        a) Nesting behaviour 2
        b) Aspects of foraging 3
        c) Parasitic behaviour 3
    Phenology 4
    Higher taxa 5
    Explanation of species text 5
    A scale for assessing Kent aculeate rarity statuses 5
    Coverage and species richness 6
    Numbers of recorded species by decade 6
Chrysidoidea 7
    Dryinidae, Embolemidae and Bethylidae 7
    Chrysididae 7
        Elampinae 7
        Chrysidinae 9
        Cleptinae 12
Vespoidea 13
    Tiphiidae 13
        Tiphiinae 13
    Mutillidae 14
        Myrmosinae 14
        Mutillinae 14
    Sapygidae 14
    Formicidae 15
        Ponerinae 15
        Myrmicinae 16
        Dolichoderinae 19
        Formicinae 20
    Pompilidae 22
        Pepsinae 23
        Pompilinae 26
        Ceropalinae 29
    Vespidae 30
        Eumeninae 30
        Vespinae 33
Apoidea 36
    Sphecidae 36
        Sphecinae 36
    Crabronidae 37
        Astatinae 37
        Crabroninae 37
        Pemphredoninae 48
        Bembicinae 54
        Philanthinae 56
    Apidae 58
        Colletinae 58
        Andreninae 62
        Halictinae 75
        Melittinae 85
        Megachilinae 87
        Apinae 94
Appendix 1. Aculeate species possibly but inconclusively recorded from Kent **109**
Appendix 2. Aculeate species that might yet be recorded from Kent **109**
Appendix 3. Unusual and inconclusive aculeate specimens found in the county ... **109**
The Natural England Natural Areas of Kent **110**
References & Further Reading **111**
Index of covered taxa **113**

THE BEES, WASPS AND ANTS OF KENT

Summary

440 distribution maps and 7 composite maps of 448 Kent aculeate species are introduced here, plotted from a database of approximately 55,000 records. All families are covered apart from the Dryinidae, Embolemidae and Bethylidae (an additional 22 species), where a brief resumé is given.

Introduction

This volume emerged from the compilation of a database for Kent aculeate Hymenoptera, partly for my position as the Kent Field Club's referee and recorder for the group and partly as a target species compiler for the national Bees, Wasps & Ants Recording Society (BWARS). It was originally conceived to be a recording scheme for the Kent ants but as the records for other aculeate groups grew, so did the scope of this volume.

Acknowledgements

I must acknowledge a debt of gratitude to the late John Felton for his encouragement and identification work during my early years as a collector and recorder of aculeates. Without his assistance I might not have continued in recording and this volume is dedicated to his memory; also to my late wife Alison, who understood and patiently accepted my passion for these insects.

Other aculeate specialists have identified difficult specimens for me, including Dr Michael Archer, Mike Edwards and George Else.

♀ *Coelioxys mandibularis*

Records have come not only from my own activity but also from many other collectors' work; a complete list of contributors, past and present, follows. Some of these contributors provided a large number of records whilst others only one or two, such as modern hornet observations.

K Alexander, A E Allchin, J H Allchin, A A Allen, A M Allen, G W Allen, H A Allen, Pat Allen, Peter Allen, M Allison, J Anthony, M E Archer, E B Ashby, G Austen-Price, Lord Avebury, J S Badmin, D B Baker, B Banks, K E J Barrett, M Barry, E Bartlett, * Beaumont, I C Beavis, D Bennett, J Bennett, R B Benson, M Bishop, F Booth, A Braby, E Bradford, R C Bradley, F Brightman, J P Brock, G Brook, J Brook, A Brookman, A Brown, R G Brown, F R Browning, S Buell, D Burgess, J T Burn, J F Burton, L C Bushby, D Carr, P J Chandler, Phil Chantler, K Chanter, R Childs, B Chipperton, K Chisnall, A J Chitty, S Clancy, J A Clark, D Clay, A N Clements, L Clemons, C A Collingwood, G A Collins, G B Collins, S Connop, R Cooper, K Cork, H Cornally, M Craven, W C Crawley, R A Crowson, M Cuddeford, J Curtis, C W Dale, N Davies, S L Davies, O L Davis, M C Day, J Denton, G H L Dicker, W R Dolling, H Donisthorpe, C M Drake, C A W Duffield, N Duncan, M Edwards, H Elgar, M Ellison, A R Elton, G Elworthy, S Elworthy, G R Else, W A Ely, M Enfield, S J Falk, A Farmer, J C Felton, J Feltwell, J P Field, L Flower, D Franks, A E Fray, K Friend, G E Frisby, D C Gardner, A Golding, A S Grace, D W Grant, D Grist, S Grove, K M Guichard, E C M Haes, * Hall, A Hamm, A W Harman, C G Harris, J Harris, P R Harvey, P Harwood, R D Hawkins, G Hazlehurst, N F Heal, P Heathcote, A Heaton, G Hemington, G Hitchcock, P J Hodge, * Hope, S Hoy, M Isaacs, A P Jarman, N Jarman, A Jarrett, E A Jarzembowski, D W Jenner, M Jenner, A W Jones, R Jones, P Kirby, H Lamb, E J Langner, * Latter, J Leclerq, G LeMarchant, S Lemon, T E Lester, D F Lloyd, J Lowet, D Mailham, J R Malloch, T A Marshall, A Massee, N Matthews, A V Measday, R Mellor, S Melville, S R Miles, D Mills, J Mitchell, F D Morice, R Morris, C H Mortimer, R I Moyse, T Mullender, E Nelmes, D A Newman, G E J Nixon, * Norton, C G Nurse, H Oehl, P T Olney, A Ottley, G M Orledge, R E Oxley, L Packer, I Palmer, A Parker, M Parsons, R E F Peal, N Pearson, J F Perkins, R C L Perkins, P R Phebay, E G Philp, B J Pinchen, L Pinkham, C W Plant, * Power, S Poyser, J Puckett, * Rayward, O W Richards, S P M Roberts, R Robertson, N A Robinson, D Rolfe, V Rook, P Roper, V Rose, G A J Rothney, C Rowe, K Ruff, L Rule, A J Rundle, A Russell-Smith, C Samson, * Satchell, E Saunders, J Shorter, W E Shuckard, K C Side,

R Simpson, F W L Sladen, F Smith, R Smith, S D Smith, G M Spooner, E C Still, A E Stubbs, M Suters, F V Theobald, A Turner, D Tutt, J Tyler, A M Tynan, J Varley, R Vinall, S Wakely, A O Walker, K Wallwork, M Walter, M Waterhouse, A Watts, R C Welch, A Wells, Farren White, I Whitehouse, D Wilde, P Williams, A Wilson, J Winter, T Wishart, L H Wollatt, D Wood, C S Wood-Baker, B E Woodhams, I H H Yarrow, P F Yeo. Where an asterisk precedes a surname the initial/s are not known.

It was particularly fortunate that the notebooks of the late Dr Gerald Dicker were located by the staff of the Kent and Medway Biological Records Centre enabling this important dataset to be incorporated into the Atlas, increasing the size of the database by some 15,000 records.

I am pleased to acknowledge permission by Professor Ed Jarzembowski to examine the aculeate collections held by Maidstone Museum. These particularly provided a rich source of older records from the late nineteenth and early twentieth century collectors: H Elgar, G Frisby and H Lamb, and mid twentieth century records by K C Side and Lieutenant Colonel C A W Duffield. Professor Jarzembowski also allowed access to the Kent Biological Archives at the Museum. Some literature records have been used; particularly useful were J C Felton's works (1967, 1969) on the Kent ants. I also acknowledge the help of the staff of the Kent and Medway Biological Records Centre, who frequently passed on records and specimens for identification from the general public. They were also very supportive in the production of this Atlas. Several people have kindly allowed me to use their aculeate photographs; acknowledgements are with the images.

The coverage, coincidence and distribution maps have been drawn using the computer program DMAP for Windows written by Dr. A J Morton. The DMAP boundary file of Kent was digitised by L Clemons (outline) and R I Moyse (Natural Areas). Natural England are thanked for permission to use their Natural Areas on the maps. For an explanation of these natural areas see page 110.

Finally, thanks are due to J S Badmin, R I Moyse and S Roberts for proof-reading and refereeing this book.

Recording area
Kent as defined here is composed of the two Watsonian vice counties, East (vc15) and West Kent (vc16), which together cover the limits of the 1852 boundary of the county. Although this system may seem archaic as a recording area, it remains useful as it has not, and will not, alter with time. Thus it should allow meaningful comparisons with earlier studies and possible future atlases.

Classification of the organisms
The taxonomic classification of the aculeates used here essentially follows the checklist by G R Else in the BWARS Members Handbook (2005), except that the Crabroninae are arranged as in Melo (1999).

Three superfamilies of aculeates are recognised: Chrysidoidea, Vespoidea and Apoidea. The nomenclature of the genera and species also follows Else, except that *Hoplitis* (*Anthocopa*) *sensu* Else is provisionally regarded as a synonym of *Osmia* (Else, *pers. comm.*), and *Andrena scotica* is considered a junior synonym of *A. carantonica* (S P M Roberts *pers. comm.*).

The superfamily name Vespoidea is used in the broad sense of Brothers (1975), Gauld & Bolton (1988, 1996) and Else (2005), and includes the former superfamilies Scolioidea, Formicoidea, Pompiloidea and Vespoidea of Richards (e.g. 1980).

The Apoidea comprises the families Sphecidae (*sens. str.*), Crabronidae and Apidae. This classification also follows Else (2005) and is used in anticipation of a revision to the RES Handbook of Richards (1980).

A Resumé of Life Histories.

a) Nesting behaviour
Most bee and wasp species are "solitary", each female forming its own nest, provisioning this to feed its young, i.e. the larvae. A nest often contains more than one cell, depending on species, each having one egg laid in it. The male aculeates take no part in the nesting process. Having performed their sole function, they die leaving the females to do the work.

Many solitary aculeates nest in the ground, digging a burrow from which branch side tunnels, each leading to a cell. These ground nesters are often called "digger wasps" and "mining bees". In mining bees the nest cells are often water-proofed with secretions produced by the female. Other solitary aculeates are "aerial nesters", i.e. nesting above ground. The nests here may be in beetle exit burrows from dead wood, in broken pithy stems or occasionally, empty snail shells. A few aculeates nest in the soft mortar of old walls or in cracks or nail holes in brickwork. This habit may have originated from nesting in sandstone or clay cliff faces and is somewhat intermediate between ground and aerial nesting.

Most short-tongued bees (*Colletes*, Andreninae, Halictinae and Melittinae) are mining bees whilst the long-tongued Megachilinae are frequently aerial nesters. Some *Hylaeus* (Colletinae, short-tongued) are wall and cliff face nesters, whilst others are true aerial nesters.

Some mud-daubing aculeates construct nests in sheltered places consisting of cells entirely formed from mud mixed with salivary secretions. The paste dries like cement, with the brood developing in the cells. Mud-daubers include *Ancistrocus oviventris* (Vespidae) and *Osmia rufa* (Apidae), although the latter also exploits a range of preformed cavities.

A development of solitary nesting is found in the communal forms. Here, as in some species of *Andrena* (Apidae) and *Crossocerus* (Crabronidae), the females sometimes share a common nest entrance tunnel and each female is believed to construct her own nest burrow off this. The coming and going of the foraging females may help to deter parasites.

Some *Halictus* and *Lasioglossum* (Apidae) have become primitively eusocial, and are discussed further under the section on phenology. In eusocial species, two or more females share a nest, there being a reproductive division of labour, where one individual does most of the egg laying whilst the others mainly do the work of the colony i.e. foraging, nest construction and brood rearing. Also there has to be an overlap of generations, such that daughters help mothers during the life cycle. In our native social wasps and bumblebees, only the queen survives the winter, having mated the previous autumn, and she founds a colony during a solitary phase in the life cycle. The first brood reared in the nest are all small females called workers. These take over the running of the colony and the foundress or queen devotes most of her time to egg laying. Later in the colony life cycle males and sexual females are reared. These mate and the females go into hibernation until the following spring. In the honeybee *Apis mellifera*, a non-native, the queen is never on her own; the whole colony of queen and her workers survives the winter, feeding on stored honey, and colony reproduction is by swarming, sometimes called "fission". Here the old queen departs the nest or hive with a proportion of the workers and a new colony is established. The colony fragment remaining produces a new queen from developing queen brood. In ants the colony life cycle most often takes several years but there is usually a solitary queen phase in the cycle. However, some ant species adopt newly mated queens of their own species and the colonies are then potentially immortal.

b) Aspects of foraging
Most solitary aculeates are very specific in their choice of food for the larvae and also where they gain the carbohydrates to power their flight. Solitary wasps often prey on a very narrow range of species, e.g. from one Order or one taxonomic family. In some species the female habitually provisions only one prey item per cell whilst in others, several smaller prey are used. Pompilidae use only one spider per cell; the size of the resulting adult wasp is entirely dependent on the size of the prey and can vary quite considerably in some species.

Many solitary bee species collect pollen (as larval food) from a restricted range of plants. For example, *Heriades truncorum* forages on a small range of yellow flowered Asteraceae, including *Senecio jacobaea*, to provision its cells. The adult female bee may visit a wider range of flowers for nectar only. Short-tongued bees tend to visit flowers with a short corolla, as do many solitary wasps. Good examples of such flowers are Rosaceae, Asteraceae and umbellifers (Apiaceae). Long-tongued bees e.g. bumblebees, can use flowers with long corollas such as *Digitalis* and Lamiaceae.

c) Parasitic behaviour
The most primitive aculeates are often "parasitoids", i.e. the larva feeds on a host insect or spider left *in situ*, eventually killing it. (True parasites do not kill their host). Parasitoid behaviour is most often noted for the large group of hymenopterous superfamilies formerly classified as the Parasitica. Many an enthusiatic lepidopterist has found a large, interesting caterpillar in the field and collected it, hoping to rear a beautiful moth or butterfly, only to have their hopes dashed when a parasitoid ichneumonid wasp emerges from the cocoon.

The life cycle of some aculeate wasps is very similar. The female parasitoid aculeate has a morphologically defined sting apparatus, which it uses to subdue the host and then lays an egg. This hatches into a larva that feeds externally on the paralysed insect. Parasitoid aculeates leave the host *in situ*. Sometimes it recovers enough to continue feeding, carrying the parasitoid larva with it. Some species of *Arachnospila* (Pompilidae) behave like this, whilst others in the same genus dig a burrow in which to place the prey and then lay the egg; this latter behaviour is then predation.

The rubytailed wasps (sometimes called jewel or cuckoo wasps, family Chrysididae) are classified as parasitoids but enter the nest of the host (another aculeate species) to lay an egg in the host brood cell. The young rubytail larva which hatches waits for that of the host to finish feeding and reach the pre-pupal stage before commencing to feed on it. This is sometimes called "brood parasitism". Mutillid wasps behave similarly but the females are wingless.

In the "cleptoparasites", sometimes known as "nest parasites" or cuckoos, the larva does not feed on the host brood but on the provisions stored by the host female for that of her own. Most cleptoparasites are aculeates. The most interesting case is that of the sapygid wasps, the larvae of which feed on pollen stored by the host bee. Some bees are cleptoparasites, usually placed in genera phylogenetically close to their hosts. Thus *Sphecodes* is closely related to *Halictus* and *Lasioglossum*, whilst *Stelis* is close to *Anthidium* and *Osmia* (*sens. lat.*). Other cleptoparasitic bees have radiated to less closely related hosts, e.g. *Nomada* is usually found on *Andrena* and *Epeolus* on *Colletes*. Most interesting is *Coelioxys*, often found on the closely related *Megachile* but also the same species sometimes on the more distant *Anthophora*. In the less advanced cleptoparasites, the adult female destroys the host egg or young larva, whilst in more advanced cases, the parasitic larva may have grotesquely enlarged mandibles in the first or second instar in order to carry out the assassination.

Nest parasitism can also include the "social parasitism" exhibited by some species of *Bombus*, formerly placed in the genus *Psithyrus*. The female cuckoo *Bombus* emerges in the spring a little later than its host. The host queen will have formed a small, incipient colony with several workers and some developing brood when the parasite invades. If the nest is too large, the parasite may be repelled or killed whilst if too small, the number of parasites that can be reared will be correspondingly small. Thus timing can be quite critical for success. The usurping female, if she gains access, hides in the nest material until she has acquired the nest odour (the badge of recognition of nest mates) and then is free to move around the colony. She becomes the alpha female, either killing the host queen or deposing her by intimidation and becomes the sole egg layer. The resulting offspring are only males and females, there being no parasitic worker caste. This dependency is sometimes called inquilinism. A similar phenomenon occurs in the social wasps, although the sole British nest parasite, *Vespula austriaca*, has not so far been recorded from Kent.

When one turns to the ants, there is a multitude of forms of social parasitism. The nests of some species of ants habitually adopt queens of their own species, which have lost the ability to found nests independently. These queens may sometimes seek and achieve adoption in the nests of closely related species, the workers of which rear the alien worker brood. These are semi-social or temporary parasites. The host queen is eliminated and eventually the nest contains only the semi-social parasite queen and her workers; these latter have all of the abilities needed to look after the brood and forage etc. An evolutionary development of this is the reduction of the social parasite's worker caste, both in numbers and vigour, the species then slipping into inquilinism, where it is dependent on the host workers throughout the life cycle including the rearing of the sexual brood. Eventually, the worker caste disappears in evolution, leaving a degenerate inquiline species.

An interesting parallel is that of the slave raiders. There are only two British species in this category. *Formica sanguinea* is in an early stage of development, fully capable of living without slaves (also known as auxiliaries), whilst the non-Kentish *Strongylognathus testaceus* is very degenerate, with the host queen *Tetramorium caespitum* surviving in the nest and continuing to produce a supply of "slave" workers. *S. testaceus* is only one step removed from being an inquiline.

Another interesting symbiosis is "xenobiosis", where a small species of ant lives in the nest of another, but larger, species of ant, the broods of the two species being kept separate. This type of relationship is probably parasitic, although this is not conclusively proved. The xenobiotic ant workers move freely through the nest of the host, rarely being challenged. The only British and Kent species is the rarely recorded *Formicoxenus nitidulus*, found in the nests of *Formica* wood ants.

Phenology

It has been possible to produce descriptive comments on phenology for some species, based on frequency of recording by week of the year. Generally, the more records there are for a species, the clearer the phenology becomes. Most solitary aculeates have a *unimodal* frequency curve, where males appear a few days before females; the numbers peak at a maximum frequency with the emergence of the females and then tail off. These species are *univoltine*, i.e. have only one generation per year. In *strongly unimodal* species, the peak comes not long after emergence of the first specimens and the subsequent decline in frequency of capture is quite rapid. However, in some species, e.g. *Andrena nigroaenea*, there is a staggered emergence, so that although males can be captured in mid- to late March, fresh females of the same generation can be found in July, foraging on into late August. Such species are often *weakly unimodal*.

In some solitary species, the frequency of occurence is *bimodal*, i.e. with two peaks. Often these are the two generations of the same species, which is described as *bivoltine*. The peaks may be completely separated, defined as *strongly bimodal*, or may grade to some extent. In some bivoltine bees, the two generations may differ very slightly in morphology, such as in the sculpturing or puncturation of the cuticle, or in colour. In a few species, two frequency modes may broadly overlap and here it may be the case that the males emerge well before the females, to feed or establish territories before coming into breeding condition.

Some species of *Halictus* and *Lasioglossum* (Apidae) have a bimodal frequency with overlapping modes. Here, though, there is only one generation per year: the males and females mate in the autumn and the latter go into hibernation, emerging the following spring to nest and rear their brood. This new generation often appears before the old females have died and the result is overlapping peaks in frequency. However, another interesting halictine life cycle can look very similar: the spring females rear a first brood largely or solely of females which become workers, subservient to the dominant colony foundress or gyne, and these enlarge the nest and forage for a new sexual brood. In one species, *Lasioglossum malachurum*, there are two successive broods of the morphologically distinct workers, the second larger in mean size, before the males and largest females are produced at the end of the season. These mate and the females over-winter, often in the maternal nest.

In the cleptoparasitic aculeates, some species have a life cycle strictly defined by a single host. *Nomada flava* is strongly unimodal, as its host *Andrena carantonica* (= *scotica*). *Andrena carantonica* is also the usual host of *Nomada marshamella* but this latter species is bimodal, having a small summer generation when *A. carantonica* is virtually absent. The host here may be the closely related, but scarcer, *A. trimmerana*, a bivoltine species. A few *Nomada* have several hosts, each with differing phenologies and consequently the observed frequency of capture does not elucidate the life cycle.

Explanation of the distribution map symbols

Four date class periods, each represented by its own symbol, are used on the maps. I had decided early in the study to make the "modern period" from 1980 onwards but this was revised to 1985. Rarity status is assessed primarily from this date class. A record from this period is represented by a black circular dot on the relevant map and this class of dot will "over lay" any older date class dots from the same tetrad. Records from 1950 - 1984 are given as green dots, and those from 1910 - 1949, yellow dots. Green dots will over lay yellow dots in the same manner. One older date class, with the least priority, pre-1910, has been defined and is represented by crosses on the maps. There is relatively little information from this period. The localities of some of the older records were not precisely defined by their recorders and so many of the dots from the latter two date classes are only approximately placed on the grid. They do, however, give some indication of relative abundance of the species during this time period and so have not been totally discarded. Finally, there have been occasions where it has not been possible to verify an unusual record or the identity of a voucher specimen has been open to interpretation. In such instances I have given the record "unknown" status, represented by a small black circle inside a larger yellow circle.

A reason for making the most recent date class symbol change from 1985, rather than a slightly older 1970 date change, was to make the maps a little more sensitive to modern changes in range and frequency. This may enable the flagging up of some species in early decline.

Higher taxa

Before the species of each family, or where appropriate, subfamily or tribe, are mapped I have attempted to give basic bionomic and taxonomic information on each of these higher taxa.

Explanation of species text

In the text below each distribution map are a few comments to put the species into a Kent context. Following these comments is an indication of the frequency of occurrence of each aculeate species, in numerical form. These numerals are as follows: the number of black tetrad dots, the total number of tetrad dots and crosses; then the recorded months in Roman numerals when the species is active in the county and finally, the recorded year range. Last mentioned are any National (Shirt, 1987; Falk, 1991) and County (Waite, 2000) statuses, and my own estimate of county status where appropriate. These statuses have been included here because at the time of production of the Kent Red Data Book (Waite, 2000) there were not enough data to provide useful criteria for assessment.

A scale for assessing Kent aculeate rarity statuses

Having decided that statuses should be derived principally from modern records, a scale has been devised here for status assessment. The lower case "p" before each status stands for "proposed". As some of the 20th century recording was done at the 2km square (tetrad) level, this was deemed the best scale of quantity for the purpose. Hence, a species rarity status defined in this book is derived principally from the number of tetrad dots dated from 1985 to 2007 on its distribution map:

- 0 modern tetrad dots but some pre-1950
    pKRDB1+ (possibly extinct in the county)

- 0 modern tetrad dots but one/some in category 1950-84 *or* 1 or 2 modern tetrad dots *or* both of these alternatives
    pKRDB1 (endangered)

- 3 - 4 modern tetrad dots
    pKRDB2 (vulnerable to extinction)

- 5 - 8 modern tetrad dots   pKRDB3 (rare)

- 9 - 15 modern tetrad dots   pKa (Kent Scarce A)

- 16 - 24 modern tetrad dots   pKb (Kent Scarce B)

- 25 - 40 modern tetrad dots
    local (if clustered) or widespread (if not clustered)

- 41 - 80 modern tetrad dots   common

- 81 & over modern tetrad dots   abundant

- A further category is given, for rare species of unknown status in the county: pKRDBK

This scale has been quantified to enable most species to be put in a Kent category at least nearly equivalent to their National status. It has not been used rigidly, however. Some flexibility has been found necessary where a species, genus or even family is likely to be under- or *relatively* over represented in collections. For example, species in the Chrysididae and Pompilidae can be difficult to capture and so may appear scarcer than they actually are. Some scarce species, when their jizz and microhabitat parameters had been recognised, were deliberately sought and found. Whilst such species cannot be said to be over recorded (as this cannot be true), the resulting data are skewed compared to related, sometimes more common, species. A similar phenomenon has happened with the ants, where digitising Felton's papers has disproportionately increased the corresponding number of green dots per species map - unfortunately giving an artificial impression when perusing the maps of a declining ant fauna.

The Bees, Wasps and Ants of Kent

Coverage and species richness

Coverage

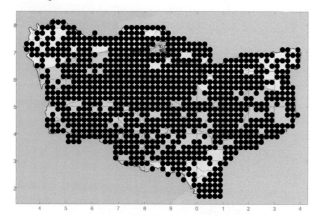

This coverage map shows the total of all known aculeate sites in Kent, from the earliest recorders in the nineteenth century to the present.

There are aculeate records from 898 of 1121 tetrads covered by this atlas.

Species richness

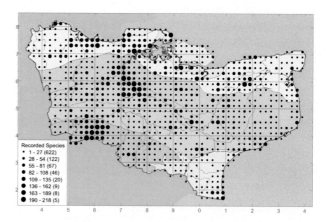

Unfortunately, this map merely shows where recorders have been most active, due to rather incomplete recording. Hence, it does not necessarily give an indication of species richness per tetrad.

Because the recording for the Atlas is incomplete, the distribution maps should be viewed bearing this map in mind. It is possible that some rarities have only been found where extensive recording has been carried out and hence may be more widespread than realised at present.

Numbers of Recorded Species by Decade

Decade	Recorded Species in Vice County 15	Recorded Species in Vice County 16	Recorded Species in both Vice Counties
Pre-1900	284	235	332
1900-1909	145	156	205
1910-1919	25	11	31
1920-1929	25	47	64
1930-1939	15	29	40
1940-1949	11	41	49
1950-1959	84	107	150
1960-1969	175	125	216
1970-1979	311	278	343
1980-1989	330	359	381
1990-1999	305	315	358
2000-2007	301	341	372
Total	426	408	448

These species totals are necessarily approximate, as they include complexes where, in past decades, the individual component species had not been separated out. This means that the complexes may be counted as one species each and the true number might not be known.

Superfamily Chrysidoidea

This superfamily has four British families, as given below.

Families Dryinidae, Embolemidae and Bethylidae

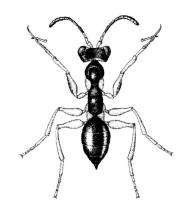

♀ *Gonatopus distinctus*

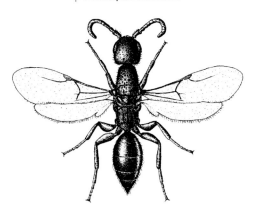

♀ *Epyris niger*

It has not been possible to produce an account with meaningful distribution maps for the species of these families. They are obscure, small to tiny, often black, species with life cycles more like the parasitoid superfamilies than aculeates. However, the morphologically defined sting apparatus of the females places these insects in the aculeates. They are most frequently found by sweeping or by trapping, and hence not often captured by aculeate collectors using more conventional methods. Some species have wingless or brachypterous females and a few are to be found in ant nests.

The life histories of this group of species are best described as parasitoid. The female stings the host insect, paralysing it and lays one to several eggs on its surface. These hatch into larvae that feed externally. The host recovers from paralysis and continues to feed, but eventually is killed. Dryinidae and Embolemidae are usually parasitoids of leaf hoppers (Hemiptera: Auchenorrhyncha, particularly Cicadellidae), whilst Bethylidae mainly parasitise coleopterous or lepidopterous larvae.

The species so far recorded from Kent are, Dryinidae: *Aphelopus atratus, A. melaleucus, Anteon arcuatum, A. exiguum, A. flavicorne, A. tripartitum, Lonchodryinus ruficornis, Gonatopus bicolor, G. clavipes, G. distinctus, G. lunatus, G. striatus*; Embolemidae: *Embolemus ruddii*; Bethylidae: *Cephalonomia formiciformis, Epyris bilineatus, E. niger, Pseudisobrachium subcyaneum, Goniozus claripennis, Bethylus boops, B. cephalotes, B. dendrophilus* and *B. fuscicornis*. This list is thought to be very incomplete and unrepresentative; only three species have been regularly recorded, the bethylids *Epyris niger, Bethylus cephalotes* and *B. fuscicornis*. Other species may be present in collections but await determination by someone more familiar with these insects.

This account, however unsatisfactory, is the best that can be produced for the present and a more detailed treatment of the Kent species will have to wait, possibly for another author.

Family Chrysididae

This family comprises the rubytailed wasps and their relatives. They are very active insects that usually have bright metallic colours. Some species are common although frequently overlooked by the lay person. Identification can prove difficult and is often only accomplished by the expert. Unfortunately, there has been much disagreement about species distinctions in some genera and the *Chrysis ignita* complex has been treated very differently by several modern authorities. Some pinned specimens can have as many as three different species names on multiple det. labels!

Three subfamilies have been recognised for the British Chrysididae and these all occur in Kent. The species with known life cycles are parasitoids or less often cleptoparasites, and in most cases the hosts are other aculeates.

Subfamily Elampinae

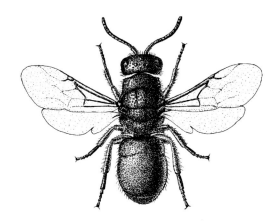

♂ *Hedychridium ardens*

Elampinae are mostly smaller in size than Chrysidinae and tend to be more compact with a shorter gaster. Like Chrysidinae the gaster is concave beneath. The hosts of most species are crabronid wasps. The *Omalus* group of genera tend to parasitise Pemphredonini, particularly twig- and post-nesting forms. *Hedychrum* species parasitise Philanthini, *H. niemelai* on *Cerceris* and a non-Kentish species, *H. rutilans*, on *Philanthus*.

Omalus aeneus (Fabricius)

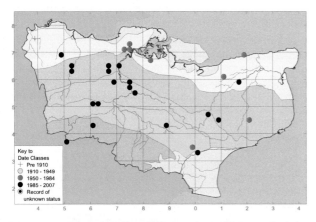

This is a rather scarce but widespread species, records being scattered across most soil types in the county. It is a parasitoid of twig- and post-nesting Pemphredonini.

Occurrence: 19, 28 *vi-viii* 1899-2004
No status.

Omalus puncticollis (Mocsary)

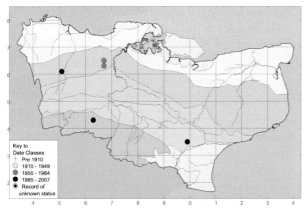

A rare species but must sometimes be confused with the preceding. Hosts are unknown but probably similar to *O. aeneus*.

Occurrence: 3, 5 *viii* 1983-2005
National status Shirt: RDB3 Falk: pNa
Kent status Waite: Notable Present work: pKRDB3

Pseudomalus auratus (Linnaeus)

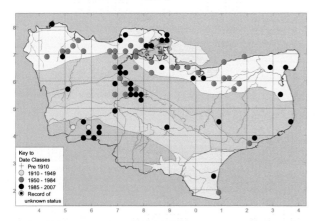

The most frequently recorded elampine in the county but apparently scarce on clay soils. Often found in gardens and a species that is likely to be under recorded there. Hosts are post- and twig-nesting Pemphredonini, particularly *Pemphredon* and *Passaloecus*, but also other Crabronidae such as *Rhopalum* and *Trypoxylon*.

Occurrence: 38, 77 *iv-ix* 1896-2007
No status.

Pseudomalus violaceus (Scopoli)

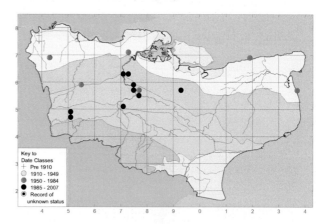

A very scarce species in the county but found on various soil types. Has been recorded from gardens. Hosts are listed as *Pemphredon* and *Passaloecus*.

Occurrence: 8, 16 *v-ix* 1898-2007
National status Falk: Nb
Kent status Waite: Notable Present work: pKa

Elampus panzeri (Fabricius)

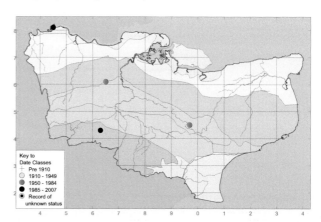

A rare species in Kent but possibly under recorded. There is apparently a rare window of opportunity to record the species, particularly males, early in the morning in grassy areas on sandy heath. British hosts are recorded as the ground nesting *Mimesa* spp (Psenini) although twig-nesting Pemphredonini may be used on the continent.

Occurrence: 2, 4 *vi-vii* 1963-2006
Kent status Present work: pKRDB3

Hedychridium ardens (Latreille in Coquebert)

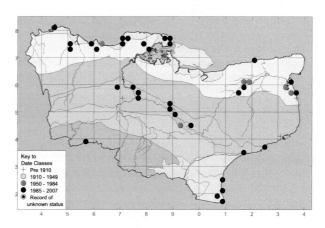

Elampinae to Chrysidinae

The distribution of this species in the county is probably limited by the scarcity of sandy heaths, its main habitat. It is common on heaths and coastal dunes, but otherwise only found on the London clay, where it is sporadic. The recorded host is the ground nesting crabronid, *Tachysphex pompiliformis*.

Occurrence: 33, 42   v-viii   1897-2006
No status.

Hedychridium cupreum (Dahlbom)

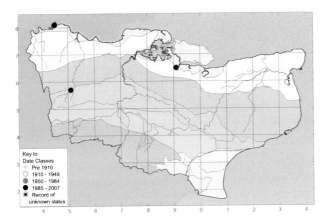

A rare species, recorded only in two consecutive years in the county. On sandy soils - alluvial and lower greensand. The host is *Dryudella pinguis* (Astatinae).

Occurrence: 3, 3   vi-vii   1996-1997
National status   Falk: Nb
Kent status   Waite: Notable   Present work: pKRDB2

Hedychridium roseum (Rossius)

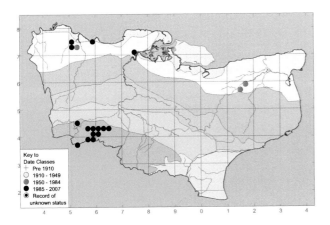

A scarce species confined to the sands in the county but apparently not recorded from the lower greensand. Hosted by *Astata boops*. This may be the only British host but in continental Europe *Tachysphex pompiliformis* and *Harpactus tumidus* may also be used.

Occurrence: 15, 18   vii-ix   1984-2007
Kent status   Present work: pKb

Hedychrum niemelai Linsenmaier
Synonymy: Sometimes misidentified as *H. nobile* or *H. aureicolle*.

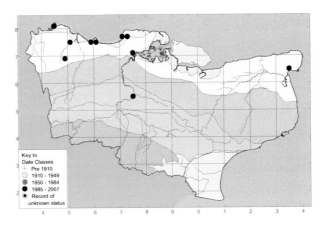

Scarce and mainly from Vice County 16, on Eocene and alluvial sands. Hosts are various species of *Cerceris* (Crabronidae; Philanthinae). The range of the *Hedychrum* in Kent particularly reflects that of *C. quinquefasciata*.

Occurrence: 10, 10   vi-viii   1983-2007
National status   Falk: pRDB3
Kent status   Waite: KRDB2   Present work: pKa

Subfamily Chrysidinae

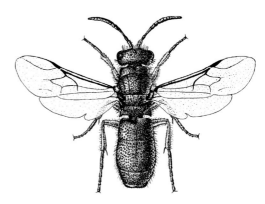

♂ *Chrysura radians*

This subfamily comprises the rubytails in the strict sense, but not all species have the characteristic metallic red gaster, *Trichrysis cyanea* having a more uniform blue-green body. The species tend to be more elongate than elampines and many have teeth at the tip of the gaster, which can sometimes provide characters for identification.

In the *Chrysis ignita* complex, different authors have had very different interpretations of the species. In this volume, the names of Morgan (1984) are followed, as understood by Dr M E Archer. I have attempted to use only records for species of the *C. ignita* complex that have been confirmed from specimens identified either by Archer or M Edwards. Nevertheless, the maps are sufficiently unreliable that a composite map of the complex is included.

The species are parasitoids or cleptoparasites of other aculeates, many species being found on eumenine wasps. *Trichrysis cyanea* is believed to parasitise various wood nesting crabronid wasps, including *Trypoxylon* spp., and some *C. ignita* complex species will use crabronids as alternative hosts. *Chrysura radians* is found on osmiine bees.

Trichrysis cyanea (Linnaeus)

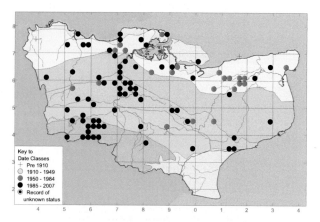

An abundant species widely spread across the county. Found on most soil types but most frequently recorded from on the sands. It parasitizes wood-nesting crabronids, particularly *Trypoxylon* spp. but also *Pemphredon* spp. An unusual host is the osmiine bee, *Chelostoma florisomne*.

Occurrence: 67, 91 v-x 1896-2007
No status.

Chrysis angustula Schenck
Synonymy: *C. gracilis*. A *C. ignita* complex species.

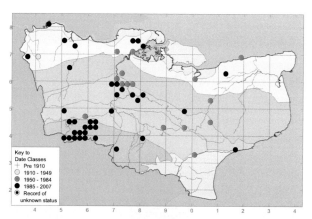

Particularly frequent on the sand but found on most soil types. Probably absent from Channel coastal marshes. Hosts are eumenine wasps, perhaps most often *Ancistrocerus trifasciatus*. Also recorded as hosts are larger crabronines.

Occurrence: 38, 54 iv-ix 1931-2007
No status.

Chrysis fulgida Linnaeus

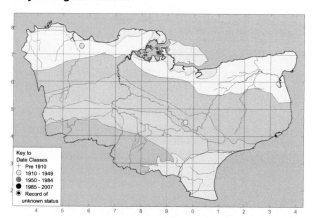

Assumed extinct in the county, although recently rediscovered in Surrey and Dorset (D W Baldock and S P M Roberts respectively, *pers. comm.*). The only reliably recorded British host is *Symmorphus crassicornis* (Vespidae; Eumeninae).

Occurrence: 0, 2 v-vii 1946-1949
National status Shirt: RDB1 Falk: pRDB1
Kent status Waite: KRDBK Present work: pKRDB1+

Chrysis gracillima Foerster
Synonymy: *Chrysogona gracillima*.

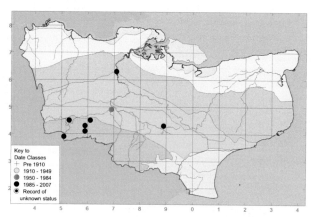

First recorded as a British species from Kent. So far found mainly south of the lower greensand in the county but may be spreading. The hosts are unknown but may include *Trypoxylon clavicerum* (Crabronidae).

Occurrence: 7, 8 vi-viii 1977-2005
National status Shirt: RDB2 Falk: pRDB2
Kent status Waite: KRDB1 Present work: pKRDB3

Chrysis ignita aggregate

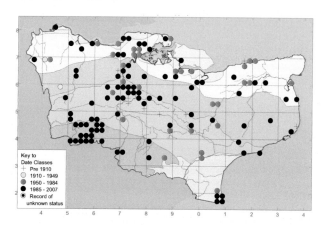

A composite map including *C. angustula*, *C. ignita*, *C. impressa*, *C. longula*, *C. mediata*, *C. rudii*, *C. rutiliventris* and *C. schencki* and unplaced records. This complex includes some of the most frequently recorded chrysidids in the county.

Chrysis ignita (Linnaeus)
A *C. ignita* complex species.

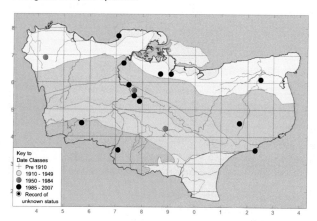

Frequent on chalk and sands, but scarce on clay soils. Probably absent from Channel coastal marshes. A parasitoid of *Ancistrocerus* species; noted in the literature (Morgan, 1984) are *A. parietum* and *A. scoticus*. The dot map however, suggests that *A. gazella* is the principal host in Kent. Some older host records may be based on misidentifications of the *Chrysis*.

Occurrence: 12, 15   v-x   1895-2007
No status – firm identifications are scarce and there has been possible confusion with other species at times.

Chrysis illigeri (Wesmael)
Synonymy: *C. helleni*

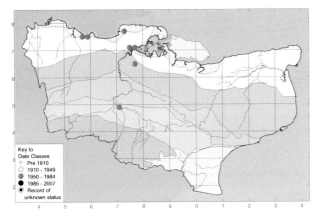

A very scarce species formerly found principally on sands and the chalk. It was recorded in the county for only a short period and there are no modern data. The recorded host is *Tachysphex pompiliformis*.

Occurrence: 0, 7   v-viii   1980-1983
National status   Falk: Nb
Kent status   Waite: Notable   Present work: pKRDB3

Chrysis impressa Schenck
A *C. ignita* complex species.

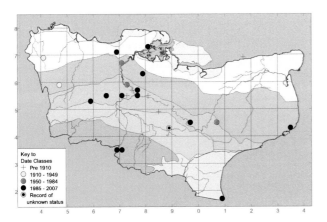

It is my belief that this species is very under recorded, perhaps partly through confusion with other *C. ignita* complex species. Widespread and most frequent on sandy soils. Recorded hosts include *Ancistrocerus trifasciatus* and *A. parietum*; *A. gazella* is a putative host.

Occurrence: 13, 24   v-viii   1895-2006
No status.

Chrysidinae

Chrysis longula Abeille de Perrin
A *C. ignita* complex species.

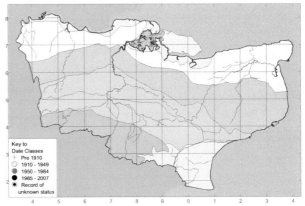

Assumed long extinct in the county; it had been found only at one site. Hosts are believed to be *Ancistrocerus antilope* and *A. parietinus*.

Occurrence: 0, 1   vi   1897
National status   Shirt: RDB3   Falk: pRDB3
Kent Status   Present work: pKRDB1+

Chrysis mediata Linsenmaier
A *C. ignita* complex species.

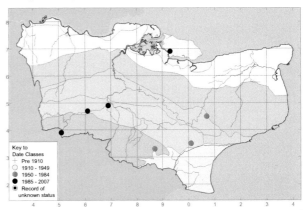

This species is probably under recorded for the same reasons as *C. impressa*, but is obviously scarcer than that species. Recorded hosts are *Ancistrocerus trifasciatus* and *Odynerus spinipes*.

Occurrence: 4, 8   vi-viii   1896-2007
Kent status   Present work: pKa

Chrysis ruddii Shuckard
A *C. ignita* complex species.

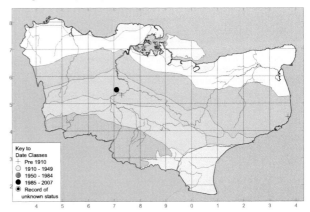

Recently refound in the county (male specimen det Dr. M E Archer but see Appendix 3). The usual host is *Ancistrocerus oviventris*, a strongly declining species that has not been recently recorded from the modern *C. ruddii* locality.

Occurrence: 1, 3   v-vi   1890s, 1998
Kent status   Present work: pKRDB1

Chrysis rutiliventris Abeille de Perrin
Synonymy: *C. vanlithi*. A *C. ignita complex species*.

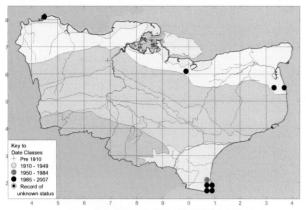

Confirmed modern records are confined to parts of the coast, where it may be locally numerous, e.g. at Dungeness. Hosts are likely to include *Ancistrocerus oviventris* and particularly *A. scoticus*.

Occurrence: 8, 10   *v-viii*   1900-2002
Kent status Present work: pKb

Chrysis schencki Linsenmaier
A *C. ignita complex species*.

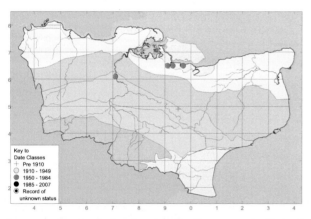

A rare species, most often found on the Eocene sands. No hosts have been recorded.

Occurrence: 0, 5   *v-viii*   1904-1983
National status Falk: Na
Kent status Waite: Notable   Present work: pKRDB2

Chrysis viridula Linnaeus

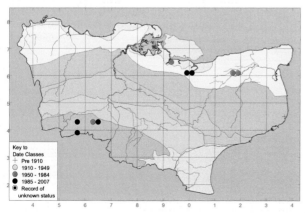

Formerly more common than at present. Recorded only from sandy habitats, with one coastal site. The hosts are *Odynerus spinipes* and *O. melanocephalus*.

Occurrence: 5, 11   *v-viii*   1897-2006
Kent status Present work: pKRDB3

Chrysura radians (Harris)
Synonymy: *Chrysis pustulosa*

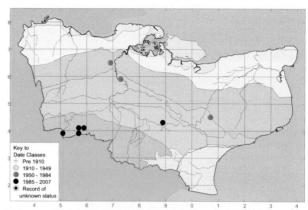

A rare species but probably under recorded. Modern occurrences are only south of the lower greensand but it was formerly more widespread. Hosts are Osmiina (Apidae), including *Osmia leaiana*.

Occurrence: 5, 9   *v-ix*   1880-2004
National status Falk: Na
Kent status Waite: Notable   Present work: pKRDB3

Pseudospinolia neglecta (Shuckard)
Synonymy: *Spintharis neglecta, Euchroeus neglectus* and *Spinolia neglecta*.

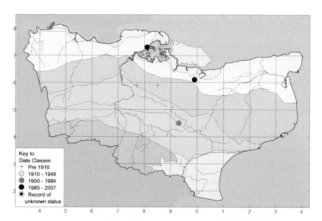

The generic position of this species is in some doubt. It is rare but found on various soil types, not especially from the sands. The dot map indicates a significant decline. The host is *Odynerus spinipes*.

Occurrence: 2, 6   *v-vii*   1899-1989
Kent status Present work: pKRDB2

Subfamily Cleptinae

The cleptines are probably the most primitive group of Chrysididae, having the underside of the gaster convex rather than concave and in this respect are unlike the other British subfamilies. There are two British species of Cleptinae, both recorded from Kent. The known hosts are nematine sawflies (Tenthredinidae), in particular the pupae. As *Nematus ribesii* has declined, so has *Cleptes semiauratus*.

Cleptes nitidulus (Fabricius)

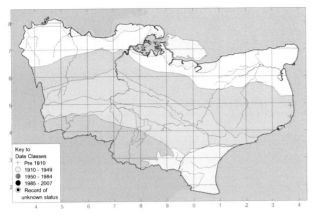

Assumed long extinct in the county, the only known records being nineteenth century. The hosts are tenthredinid sawfly pupae.

Occurrence: 0, 2   *A Kent flight period cannot be defined*   Pre-1900
National status   Shirt: RDB3   Falk: pNa
Kent status   Present work: pKRDB1+

Cleptes semiauratus (Linnaeus)
Synonymy: *C. pallipes*

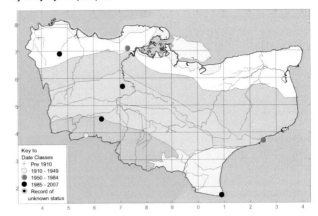

Now a rare species, sometimes coastal, with old records on various sandy soils. The recorded host is *Nematus ribesii* (Tenthredinidae).

Occurrence: 4, 9   *vii-viii*   1899-2006
National status   Falk: Na
Kent status   Waite: Notable   Present work: pKRDB3

Superfamily Vespoidea

Vespoidea is considered here to include the following British families: Tiphiidae, Mutillidae, Sapygidae, Formicidae, Pompilidae and Vespidae.

Family Tiphiidae

Abroad, this is a large, diverse family with several subfamilies. Of these only two, Methochinae and Tiphiinae, occur in Britain. *Methocha articulata* (= *ichneumonides*), the only British species of Methochinae, has so far not been found in Kent. *Tiphia* is the only tiphiine genus in Britain.

Tiphia femorata Fabricius

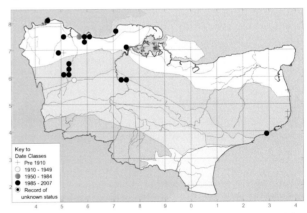

Rather scarce and mostly recorded from Vice County 16. Mainly a species of the sand but apparently on occasion found on chalk soils. Not found south of the lower greensand. A parasitoid, hosts are the larvae of scarab beetles, including *Aphodius*, *Rhizotrogus* and *Anisoplia*.

Occurrence: 15, 21   *vii-viii*   1850s-2007
No status.

Tiphia minuta Vander Linden

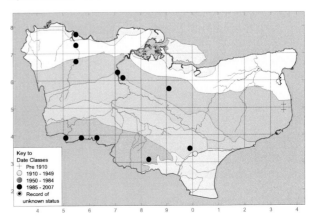

Subfamily Tiphiinae

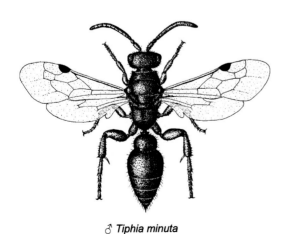

♂ *Tiphia minuta*

Britain has two species of *Tiphia* and both are recorded from Kent. Neither of these is common in the county although they can sometimes be locally numerous. *Tiphia* are parasitoids of scarabaeid beetle larvae.

A scarce species, with records scattered across the county. Due to the small size, it must sometimes be overlooked. Not found on clay soils. A parasitoid of the larvae of dung beetles.

Occurrence: 11, 14 *v-vii* 1898-2007
National status Falk: Nb
Kent status Present work: pKb

Family Mutillidae

This family, constituting the "velvet ants", is divided into two subfamilies, Myrmosinae and Mutillinae, on the basis of thoracic structure. In both of these groups the female is wingless and the winged males scarce to rare by comparison. In Mutillinae the females have a painful sting. The species are parasitoids of other aculeates.

Subfamily Myrmosinae

There is only one species in this group in Britain, *Myrmosa atra* and it is frequent in the county. The hosts are small crabronid wasps such as ground nesting *Crossocerus* spp.

Myrmosa atra Panzer

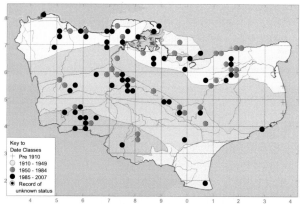

A common species but virtually confined to the sands, with some coastal sites. Hosts are recorded as both ground nesting crabronid wasps and bees, therefore it is likely to be a parasitoid.

Occurrence: 53, 92 *vi-x* 1850s-2007
No status.

Subfamily Mutillinae

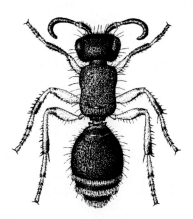

♀ *Smicromyrme rufipes*

The two British species in this group are *Mutilla europaea* and *Smicromyrme rufipes*. These have both been found in the county but are not frequent.

Mutilla europaea Linnaeus

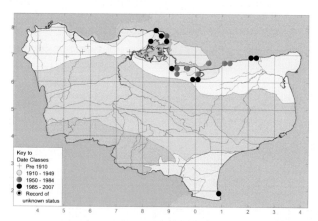

In modern terms, a scarce species of coast and estuary which reflects the current range of *Bombus muscorum*, a probable host. There are older records a little further inland. This mutillid is a parasitoid of bumblebees, *Bombus* spp.

Occurrence: 10, 22 *v-x* 1850s-2003
National status Falk: Nb
Kent status Waite: Notable Present work: pKb

Smicromyrme rufipes (Fabricius)

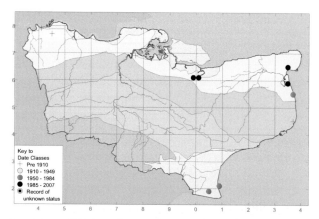

A very scarce Kent species, modern records only from Vice County 15. Found on the sand, particularly in coastal areas. The small size may mean it has been overlooked at times. A recorded parasitoid of various ground nesting crabronid and pompilid wasps, and bees.

Occurrence: 4, 11 *vi-viii* 1857-2006
National status Falk: Nb
Kent status Waite: Notable Present work: pKa

Family Sapygidae

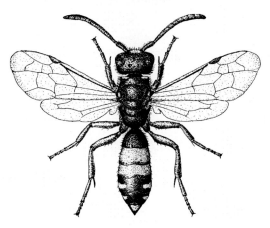

♀ *Sapyga quinquepunctata*

There are two British species in this family, both recorded from Kent. One, *Monosapyga clavicornis*, has scarcity status. Both males and females are winged. They are parasites and the hosts are osmiine bees, although some foreign species parasitise *Megachile*. It is known that the sapygid larva feeds on the nectar/pollen mass stored by the host female bee: hence, the species are considered cleptoparasites rather than parasitoids.

ingless, as worker ants can be active even in winter, if the weather is mild. I have made no attempt to map the "tramp" species – those which have been accidentally transported to this country by commerce. These include species of *Monomorium, Pheidole, Crematogaster, Linepithema, Paratrechina, Camponotus* and other genera.

Monosapyga clavicornis (Linnaeus)
Synonymy: *Sapyga clavicornis*

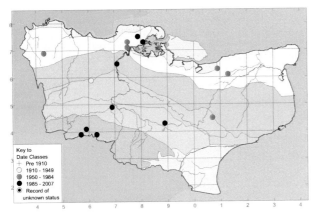

A scarce species found on various soil types and fairly frequent on the London clay. A cleptoparasite, the hosts including *Chelostoma florisomne* and *Osmia* spp.

Occurrence: 8, 15   *v-viii*   1920-2006
National status   Falk: Nb
Kent status   Waite: Notable   Present work: pKa

Sapyga quinquepunctata (Fabricius)

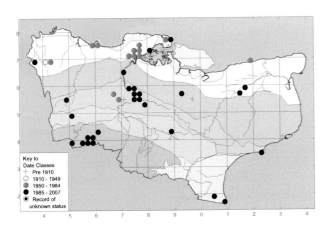

Widespread on sandy soils but with several records from the London clay. A cleptoparasite of various *Osmia* spp and *Chelostoma florisomne*.

Occurrence: 27, 44   *iv-viii*   1897-2007
No status.

Family Formicidae

The ants form a cohesive group considered a single family, Formicidae. The British species are split into four subfamilies: Ponerinae, Myrmicinae, Dolichoderinae and Formicinae. The food of many consists of scavenged insect carrion and honeydew, sometimes supplemented by nectar and the oil bodies of seeds. A few *Formica* are more active hunters of live insects, although still reliant on honeydew. The recorded activity periods for ants are relatively mean-

Subfamily Ponerinae

The Ponerinae is a distinctive group of rather elongate, cylindrical species, considered to be the most primitive British ants. There are only two species in the country, of which one might not be native. *Hypoponera punctatissima* is mainly recorded from heated buildings although records have been claimed far from human habitation. I have given it the benefit of the doubt and it is mapped here.

Hypoponera punctatissima (Roger)
Synonymy: *Ponera punctatissima*.

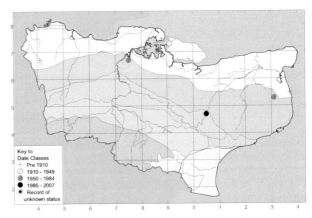

A rare species but of unknown status in the county. Most records are from heated buildings. The modern locality is Wye. Colonies are said to be "quite populous".

Occurrence: 1, 6   *vi-ix*   1886-1991
Kent status   Present work: pKRDBK

Ponera coarctata (Latreille)

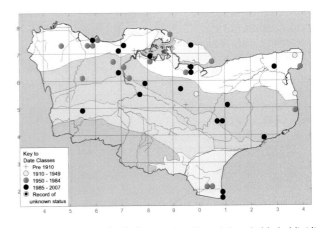

A rather scarce species in the county, although in suitable habitat it may be frequently recorded. Often it is the winged forms (alates) that are located. Found in warm, mossy situations, occasionally from gardens. Colonies often contain only a few individuals.

Occurrence: 19, 43   *iii-xi*   1860-2006
National status   Falk: Nb
No Kent status.

Subfamily Myrmicinae

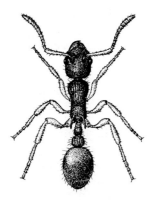

Worker *Myrmica scabrinodis*

A large, diverse group world wide, split into many tribes. The British forms are structurally rather conservative. The most speciose genus in the UK is *Myrmica*. Several members of this genus are very common in the county. Other frequently recorded myrmicines are *Leptothorax acervorum* and *Myrmecina graminicola*. Two species of social parasites (inquilines) have been found in *Myrmica* nests in Britain and these are now both treated in that genus. The inquiline *M. karavajevi* has not been recorded from the county, while there is a possible record for *M. hirsuta*, the other inquiline, and possible one for *M. vandeli*, a putative temporary social parasite. The genus *Leptothorax* has recently been partitioned and in Britain only *L. acervorum* is still placed in the genus. The other British species are now in *Temnothorax*, the oldest available name for this segregate. *Formicoxenus nitidulus* is a xenobiont in the nests of wood ants, although xenobiosis is likely to prove to be a form of parasitism. Xenobiotic ants are smaller ant species found in the nests of larger ones, keeping their brood separate in chambers of restricted size and reputedly obtaining food by cleaning the host workers. It is highly likely that some food is obtained by stealing during exchange between host workers, surely rendering this relationship parasitic. The family group name Formicoxenini has priority over the frequently used "Leptothoracini" (Bolton,1994). A very degenerate inquiline, *Anergates atratulus*, is recorded rarely in nests of *Tetramorium caespitum* in the county, the *Tetramorium* itself being rather local in Kent.

Tribe Myrmicini

Myrmica lobicornis Nylander

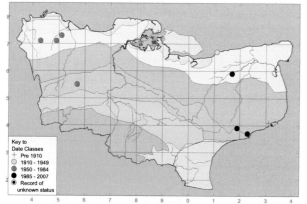

Probably under recorded but undeniably scarce in the county. A smallish, dark *Myrmica* recorded from short grass, such as pasture. Occasionally coastal.

Occurrence: 3, 9   iv-ix   1850s-1986
Kent status   Present work: pKRDB3

Myrmica rubra (Linnaeus)
Synonymy: *M. laevinodis*

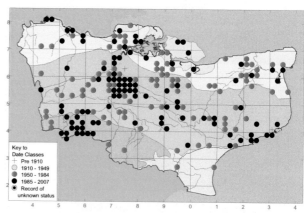

An abundant species of open places, in contrast to the following. It is frequent on lawns and in other garden situations. This species frequently visits Apiaceae and *Achillea* for nectar, and possibly preys on other insects in this situation.

Occurrence: 98, 197   i-xii   1850s-2007
No status.

Myrmica ruginodis Nylander
Synonymy: Sometimes in the past misidentified as *M. rubra*

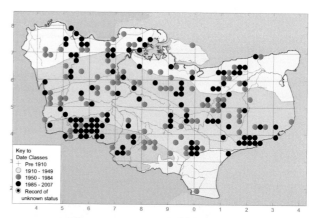

Abundant but recorded firmly in Kent only from about the time of the revision by Collingwood (1958), which established it as a good species. Particularly frequent in damp, shaded places such as woodland rides. Rare in coastal marshes.

Occurrence: 112, 212   i-xii   1954-2007
No status.

Myrmica sabuleti Meinert

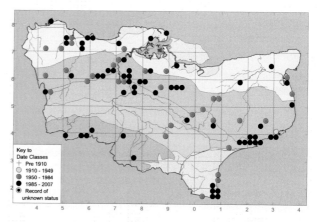

A common species very similar in morphology to *M. scabrinodis*. Recorded mainly from short turf, therefore particularly known from chalk grassland. It is also common at Dungeness, where it is found

Myrmicinae

in moss on the shingle; also from coastal sand. Sometimes found on garden lawns. Host to *M. hirsuta*.

Occurrence: 59, 92   ii-xi   1895-2007
No status.

Myrmica scabrinodis Nylander

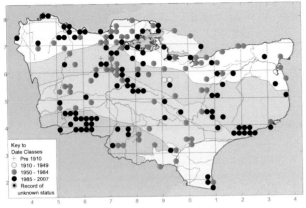

Common in many different habitats but not usually in dense woodland. Can be found on garden lawns. Several other *Myrmica* species were once regarded as varieties of *M. scabrinodis*.

Occurrence: 98, 178   i-xii   1899-2007
No status.

Myrmica schencki Viereck

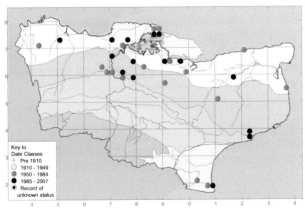

A rather scarce species but apparently a nocturnal forager. It may be under recorded – most aculeate hymenopterists are diurnal! It is recorded as being a predator of other ants. Found on the chalk and in coastal habitats, rare elsewhere.

Occurrence: 15, 32   iii-xi   1962-2003
National status   Falk: Nb
No Kent status.

Myrmica specioides Bondroit

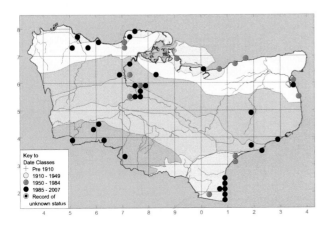

M. specioides was first recorded in Britain from the East Kent coast at Deal Sandhills in 1961. Slightly older specimens were found in a collection from Whitstable in 1958. In Kent it has been found inland since the 1970s, including on sandy soils and calcareous grassland. In the 1980s it was found on garden lawns and continues to occur in many such situations.

Occurrence: 31, 45   iii-x   1958-2007
National status Shirt: RDB3   Falk: pRDB3
No Kent status.

Tribe Formicoxenini
Formicoxenus nitidulus (Nylander)

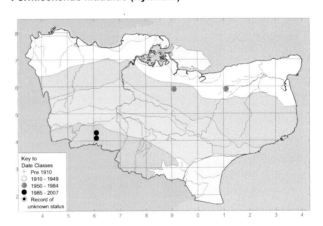

An ant living in "xenobiosis" (probably a form of inquilinism) with wood ants. It is a very small, *Leptothorax*-like species that is likely to prove very under recorded in the county. As it is only found in the nests of wood ants, in Kent the distribution will not be outside the range of *Formica rufa*.

Occurrence: 2, 4   ix-xi   1956-2003
Kent status   Present work: pKRDBK

Leptothorax acervorum (Fabricius)

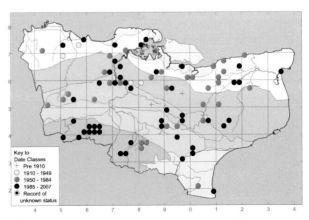

The sole British species now in *Leptothorax*, with only 11 segments to the female antenna. Widely distributed across the different soil types, including chalk, sand and clay. Not particularly found on the coast. Nests in dead wood, including tree stumps. A rather small ant that can thrive near the nests of aggressive ants like *Formica rufa* and *F. sanguinea*. Presumably, it does not have a strong odour that would give its presence away.

Occurrence: 47, 86   i-xii   1890s-2007
No status.

Temnothorax albipennis (Curtis)
Synonymy: *Leptothorax tuberointerruptus*, misidentified as *T. tuberum*.

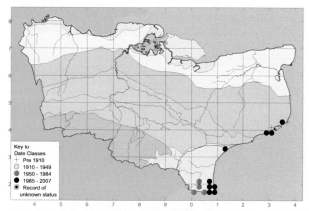

Particularly frequent on the coastal shingle at Dungeness and with other modern records only from south-east facing coasts in the county. A very small species nesting in dead gorse and broom stems, and in twigs; occasionally under stones. In some cliff habitats will nest in cracks in the rocks.

Occurrence: 9, 16   iii-viii   1852-2002
National status Shirt: RDB3   Falk: pNa
Kent status Waite: Notable   Present work: pKa

Temnothorax interruptus (Schenck)

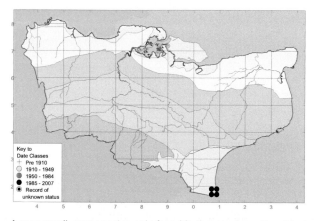

A very small, rare species only found in the county on the shingle at Dungeness. It tends to nest under stones in mossy situations but will sometimes nest in other crevices. Little is known of the habits of this species.

Occurrence: 4, 4   iv-ix   1964-1994
National status Shirt: RDB3   Falk: pRDB3
Kent status Waite: KRDB2   Present work: pKRDB3

Temnothorax nylanderi (Foerster)

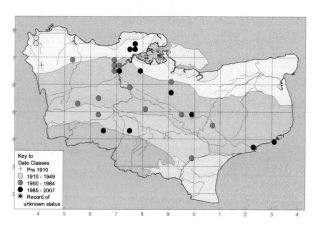

A scarce species usually found inland. The coastal records from Folkestone and Dover require confirmation. It is a very small ant often found by collectors of Coleoptera. It nests under bark, in old tree stumps and sometimes in hollow twigs on the ground. The mating flight period is in August.

Occurrence: 11, 27   i-x   1890s-2004
Kent status Present work: pKa

Tribe Tetramoriini

Anergates atratulus (Schenck)

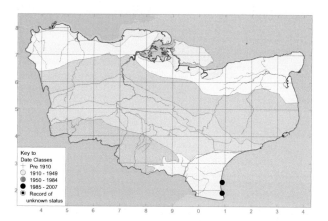

A very rare, workerless species which is an inquiline only found in the nests of *Tetramorium caespitum*. So far only two Kent occurrences are known (Dungeness, coll. R Morris; Greatstone-on-Sea, coll. G W Allen). The species is found only where there are flourishing populations of the host and is easily overlooked. It is a degenerate parasite with the male wingless and resembling a pupa. The females mate in the maternal nest and then fly to look for new host nests.

Occurrence: 2, 2   v-vi   1989-1996
National status Shirt: RDB3   Falk: pRDBK
Kent status Waite: KRDB1   Present work: pKRDB2

Tetramorium caespitum (Linnaeus)

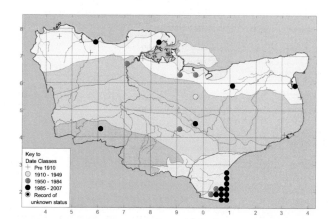

A locally common species on the Dungeness shingle. Scarce elsewhere in the county and often coastal. The species is frequent in other southern counties on dry, sandy heath. A small, dark, pugnacious ant whose mandibles open wide enough to bite human skin. Probably at least in part predacious but will take seeds and insect carrion. Mating flight period is from July to August.

Occurrence: 16, 29   iv-xi   1850-2003
Kent status Present work: pKb

Myrmicinae to Dolichoderinae

Tribe Solenopsidini

Solenopsis fugax (Latreille)

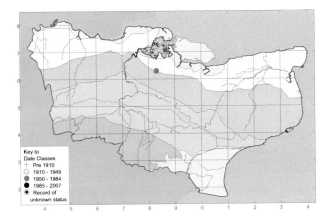

A rare, hypogaeic ant found mainly in the nests of other ants, usually *Formica* or *Lasius*. This is one of the smallest British ants and could easily be overlooked in the nest of the host. It is believed to steal the brood of the host to feed its own larvae. The queens and males are much larger than the diminutive workers. Not found recently in Kent (last record coll. J C Felton, 1964), although probably still present.

Occurrence: 0, 2   *viii*   1850s, 1964
National status   Shirt: RDB3   Falk: pRDB3
Kent status   Present work: pKRDB1

Tribe Myrmecinini

Myrmecina graminicola (Latreille)

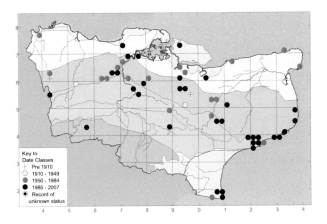

Surprisingly frequent in the county. A species of short grass, hence common on the downs and with a distribution rather similar to that of *Myrmica sabuleti*. Also frequently coastal. It is often recorded from the winged forms and again a species likely to be found by coleopterists sifting leaf litter. Another small species that feigns death when handled.

Occurrence: 33, 57   *iii-xii*   1850s-2007
No status.

Subfamily Dolichoderinae

The dolichoderines are a mainly warm temperate and tropical subfamily. The foreign genera show a variety of nesting habits, some being strictly arboricolous.

There are only two native Dolichoderinae in Britain, sibling species of *Tapinoma*. One is found in Kent, *T. erraticum*, a species which builds terrestrial nests in warm situations.

Tapinoma erraticum (Latreille)

Tribe Stenammini

Stenamma westwoodii aggregate
Including *S. debile* & *S. westwoodii* (*sens. str.*).

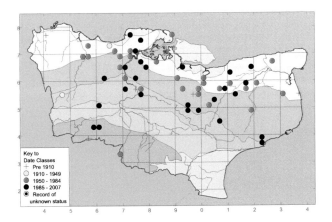

The distinction between the two species in the female castes is very unclear and the two may yet prove synonymous. Probably very under recorded and frequently found by coleopterists sifting leaf litter. Small species which feign death when threatened by more aggressive ants. They are open woodland specialists.

Occurrence: 26, 60   *i-xii*   1887-2007
No status.

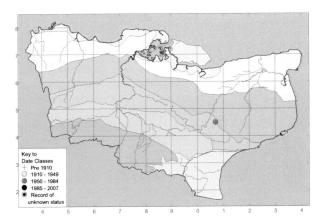

A rare species in the county with no modern data. It is important to confirm the continued existence of this, a poorly known ant in the county. An active species that nests in the ground and is known to raise solaria above ground level.

Occurrence: 0, 1   *viii-ix*   1966
National status   Falk: pNa
Kent status   Waite: Notable   Present work: pKRDB1

Subfamily Formicinae

There are two native British genera of Formicinae, *Lasius* and *Formica*. These are both Holarctic, mainly boreal taxa, and are speciose in Britain. Revisions in the late 1980s and early 1990s by B Seifert have increased the number of recognised species of *Lasius* but, at the same time, made correct identification very difficult. This is reflected in the treatment given here, where in some cases maps are of aggregates rather than individual species. Whilst this is admittedly highly unsatisfactory, there are a great many records from older sources that cannot be placed to species.

Tribe Formicini

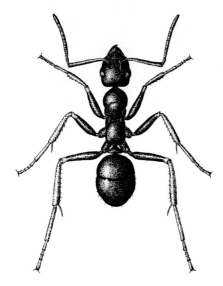

Worker *Formica fusca*

Formica cunicularia Latreille
Synonymy: *Formica fusca* var. *glebaria*, *F. fusca* var. *rubescens*

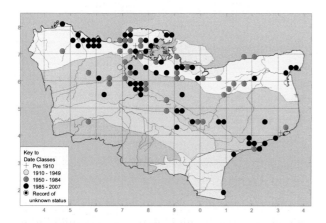

A very similar species to the next but ecologically quite different. Rather variable in colour, sometimes even nest mates vary significantly. Frequent on the coast and downs, this is a species of "hot spots". It enjoys the full strength of the sun unlike *F. fusca*, which is often found in shaded situations. A fairly large ant but not particularly aggressive. It visits umbellifers (Apiaceae) for nectar and has very good vision, which is used for hunting. Will also take honeydew.

Occurrence: 58, 103   *iii-xii*   1850s-2007
No status.

Formica fusca Linnaeus

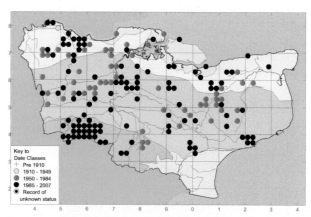

A very common species but largely absent from salt marsh and coast, where it is replaced by *F. cunicularia*. *F. fusca* is a species of woodland rides and other shaded situations. Like *F. cunicularia*, the present species will take nectar from Apiaceae and other plants, and will utilise honeydew when this is available. Occasionally host to *F. rufa* queens when these found nests as temporary parasites and also the usual victim of slave raids of the rare *F. sanguinea*.

Occurrence: 109, 167   *i-xi*   1896-2007
No status.

Formica rufa Linnaeus

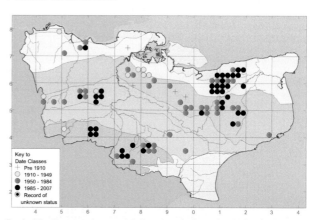

Probably the most completely recorded aculeate in the county. Although the species can teem in areas where it occurs, it has a very restricted distribution. It is a species of open, well-drained woodland, both coniferous and deciduous, and raises its conspicuous mound nests in these woods. *F. rufa* is a pugnacious ant which attempts to eliminate competitor species. It preys on caterpillars and many other insects, and usually milks aphids for honeydew. It is the sole host for *Formicoxenus nitidulus* in the county and occasionally is a temporary social parasite in the nests of *F. fusca*.

Occurrence: 39, 93   *i-xii*   1896-2006
No status.

Formica sanguinea Latreille

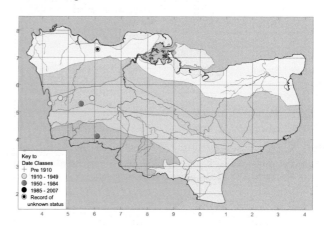

In spite of its large size and aggressive behaviour, this species has proved very elusive when searched for in the county. The nest is often in dead prostrate tree trunks. Typically an ant of dry sandy heath, this species mounts raids on the nests of F. fusca group ants and carries back to its own nest the captured brood, most of which is eaten. A small proportion survives and becomes "slaves" or auxiliaries in the F. sanguinea nest. Nests of the ant can hold rare beetles as parasites.

Occurrence: 0, 6 viii 1923-1980 (?1990)
National status Falk: pNb
Kent status Present work: pKRDB2

Tribe Lasiini

Lasius alienus aggregate
Including L. alienus (sens. str.) and L. psammophilus.

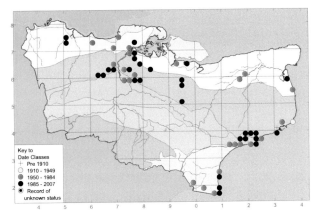

Although the two species lumped here are certainly distinct, many older data cannot be placed with any degree of certainty. Most coastal records and those from sandy heath will probably refer to L. psammophilus whilst most on the downs will be L. alienus (sens. str.) but this segregation is not 100% reliable. Given this, the species are shown here as a composite map. The ants closely resemble the garden species L. niger but do not have the erect hairs on the scape and tibiae. The species take insect carrion and tend aphids for honeydew.

Occurrence: 19, 56 ii-xi 1897-2002
No status.

Lasius brunneus (Latreille)

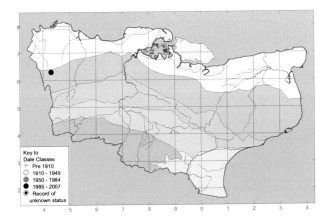

Only known by one record from typical habitat: old parkland with mature trees. It is almost certainly under recorded in the county, being an elusive, retiring species that is expanding its range. The species nests in a range of crevices in wood, including under bark in dead and living tree trunks. It subsists largely on a diet of honeydew, from aphids that are tended by the ants.

Occurrence: 1, 1 vii-viii 1999
Kent status: Present work   pKRDB3

Lasius flavus (Fabricius)
Synonymy: L. myops

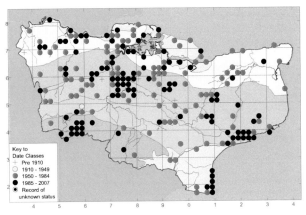

A very common and widely distributed species but the workers are not often seen above ground level except when the marriage flights occur. Found in a variety of habitats, including gardens. Where undisturbed it will raise earth mounds as solaria to warm the brood and speed up its development. The diet is probably largely honeydew from aphids tended on roots in the nest.

Occurrence: 103, 213 i-xii 1895-2007
No status.

Lasius fuliginosus (Latreille)

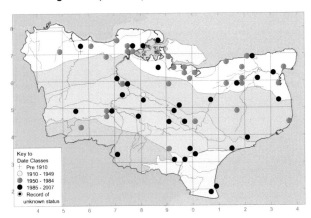

A distinctive, shining black ant. Can be locally abundant where it occurs. A temporary social parasite on ants of the Lasius umbratus group and occasionally of the L. niger group. Sometimes the origin of a colony can be seen by finding yellow workers among the black L. fuliginosus. Nests in old oak trees, where it tends aphids. Also predaceous. Often located as trails of workers moving between nest and aphid aggregations. The workers seldom deviate from these odour trails.

Occurrence: 31, 63 iii-xi 1899-2006
No status.

Lasius meridionalis (Bondroit)
Misidentified as L. rabaudi at times. A L. umbratus group ant.

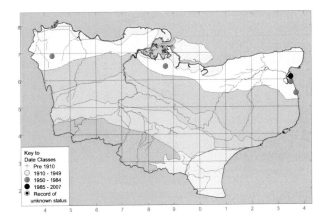

A scarce species but probably under recorded. Known only from sandy localities, so far. As with all *L. umbratus* group ants, a temporary social parasite of *L. niger* group species. Probably spends most of its life cycle underground, hence it is little known. As with other *Lasius* it is likely to be heavily reliant on honeydew secreted by aphids, which are tended by the ants.

Occurrence: 1, 5   *iv-x*   1937-1989
Kent status Present work: pKRDB3

Lasius mixtus (Nylander)
A *L. umbratus* group ant.

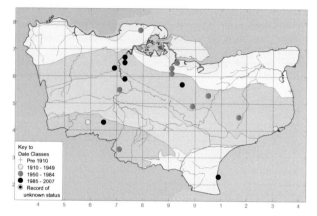

Most frequent on the chalk but also found on sandy soils, rarely coastal. Probably spends most of its life cycle underground, hence it is little known and probably under recorded. As with other *Lasius* it is likely to be heavily reliant on honeydew secreted by aphids, which are tended by the ants.

Occurrence: 7, 20   *iii-ix*   1850s-2006
Kent status Present work: pKa

Lasius niger aggregate
Including *L. niger* (*sens. str.*) and *L. platythorax*.

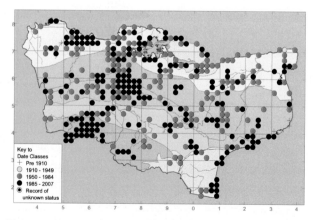

The two species forming this aggregate have gained acceptance as distinct but as there is a large body of data, much old, which does not distinguish between them, they are reluctantly mapped here as an aggregate. *L. niger* (*s.s.*) is said to be more synanthropic, whilst *L. platythorax* nests in prostrate and standing dead tree trunks, often on sandy heath and near bogs. The complex is one of the most frequently recorded aculeates in the county. The species are reliant on honeydew from tended aphids as a food source but also take live and dead insects.

Occurrence: 143, 356   *i-xii*   1850s-2007
No status.

Lasius umbratus aggregate

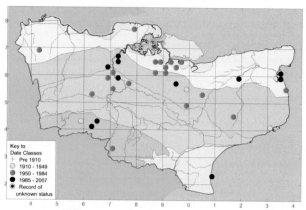

As the species in this complex are difficult to identify, the individual species maps are likely to prove less than 100% reliable. Hence the complex is mapped here. The included species are: *L. meridionalis, L. mixtus, L umbratus* and possibly *L. sabularum*. The last named species has only recently been distinguished and is not yet reported from the county.

Lasius umbratus (Nylander)
A *L. umbratus* group ant.

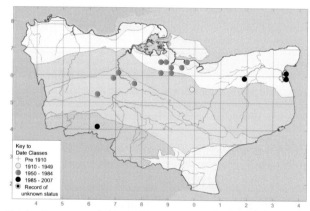

Predominantly a species of sandy soils. Probably spends most of its life cycle underground, hence it is little known and probably under recorded. As with other *Lasius* it is likely to be heavily reliant on honeydew secreted by aphids, which are tended by the ants.

Occurrence: 4, 16   *v-x*   1927-2002
Kent status Present work: pKa

Family Pompilidae

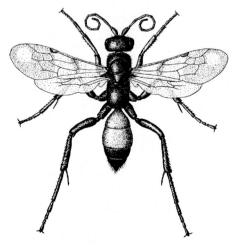

♀ *Priocnemis hyalinata*

The family Pompilidae constitutes the spider wasps, a taxon which is usually characterised by species with long legs and rather gracile bodies. They are solitary wasps that often run in an excited manner, sometimes flicking the wings and also taking short, skipping flights. They can prove difficult to capture, diving into the sward if a net is placed over them or slipping out from under the side of the net. This has meant that as a group they are under represented in collections and hence under recorded. The rather uniform habitus of the spider wasps increases the difficulty of identification, in a family in which there are taxonomically complex genera and species groups. The taxonomy and ecology of the British Pompilidae is covered by Day (1988).

The vernacular name for the pompilids is derived from their habit of hunting spiders for prey. Only one spider is used per cell and in this respect pompilids differ from the spider predators in the Crabronidae, i.e. *Trypoxylon* and *Miscophus*, which provision each cell with several small prey.

There are two British cleptoparasitic (cuckoo) genera in the Pompilidae, i.e. *Evagetes* and *Ceropales*. *Evagetes* is closely related to other pompiline genera and only separable from them with difficulty at the generic level, whilst the latter is distinctive and placed in its own subfamily, Ceropalinae. There are no known cleptoparasitic pepsines.

Many species visit umbellifers (Apiaceae) and other plants such as *Achillea* (Asteraceae) for nectar.

Subfamily Pepsinae

Although once placed near to *Dipogon* (formerly known as *Agenia*), *Auplopus* is not close to the other British pepsines and it is sometimes separated tribally from them. The paucity of old records for both Kent *Dipogon* species is not explained. The smaller species of *Priocnemis* can prove very difficult to identify and nesting is not well known for the genus. Knowledge of the distribution and prey of *Priocnemis* in Kent is largely due to the collecting of Dr G H L Dicker.

Tribe Ageniellini

Auplopus carbonarius (Scopoli)

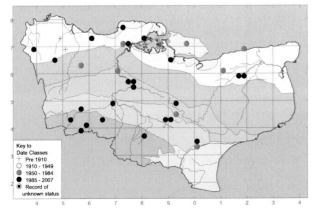

Frequently a species of open, broad leaved woodland but can also be found in shaded gardens and other habitats. Builds complete mud cells in crevices. May reuse nests of bees and other wasps, building its own cell/s inside. Preys on free-living spiders of a number of families. It sometimes disarticulates the spider before transportation to the nest.

Occurrence: 23, 33   v-ix   1890s-2007
National status   Falk: pNb
Kent status   Waite: Notable   Present work: No Kent status.

Tribe Pepsini

Dipogon subintermedius (Magretti)
Synonymy: *Dipogon nitidus*

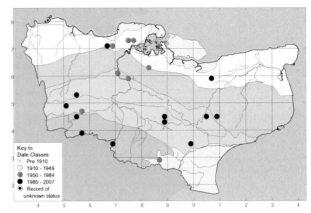

The present species is found predominantly on the sand, although it may like the scarp faces of various strata. Found in hedgerows and near dead timber. Nests in old beetle burrows, hollow stems and old walls. The prey is the spider, *Segestria senoculata* (Segestriidae).

Occurrence: 12, 21   vi-ix   1890s-2004
Kent status   Present work: pKb

Dipogon variegatus (Linnaeus)

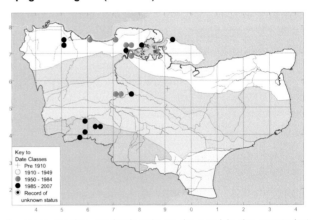

A scarce species, predominantly on the sand in the county but occasionally coastal. More restricted than the previous species. Will nest in virtually any cavity, including old walls, timber and empty snail shells. The prey is *Xysticus cristatus* (Thomisidae).

Occurrence: 11, 19   v-ix   1890s-2007
Kent status   Present work: pKb

Caliadurgus fasciatellus (Spinola)
Synonymy: *Calicurgus hyalinatus*

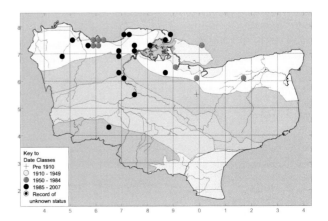

Another species with few older data. The species is particularly prevalent in the north Kent marshes, although sometimes found

24 The Bees, Wasps and Ants of Kent

inland on the sand. The female digs a short burrow in sand and provision with araneid spiders that are hunted whilst on the web.

Occurrence: 11, 13 v-x 1890s-2007
Kent status Present work: pKb

Cryptocheilus notatus (Rossius)

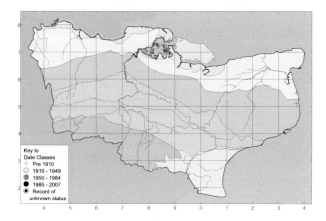

A very rare species not recorded from the county since the days of Frederick Smith (mid 19th century). A species known to excavate multi celled nests in pre-existing cavities, in a variety of sandy and grassy places, including on clay. It occurs both on the coast and in other counties, inland. The cells are provisioned with agelenid and dictynid spiders particularly, but other families may be utilised. The adults can be found on *Achillea* and Apiaceae.

Occurrence: 0, 2 A *Kent flight period cannot be given* 1858-1860
National status Shirt: RDB3 Falk: pRDB2
Kent status Present work: pKRDB1+

Priocnemis agilis (Shuckard)

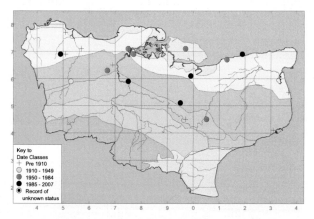

Older identifications of this species are not always reliable. It appears to occur largely on the chalk scarps and on the coast. There are no reliable prey records. The species visits *Daucus* and possibly other umbellifers (Apiaceae).

Occurrence: 5, 19 vii-ix 1850s-2001
National status Falk: pNb
Kent status Present work: pKa

Priocnemis cordivalvata Haupt

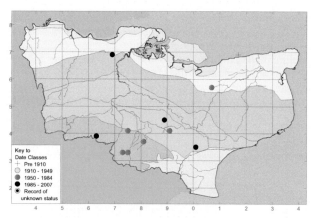

A scarce species often found in the glades and rides of mature, broad leaved woodland. Found in Kent particularly in woods on the Weald Clay and Sands. Reliable prey records are few, an immature gnaphosid spider being found carried by *P. cordivalvata* in Sussex.

Occurrence: 4, 11 vii-ix 1890s-2003
National status Shirt: RDB3 Falk: pNb
Kent status Waite: Notable Present work: pKa

Priocnemis coriacea Dahlbom

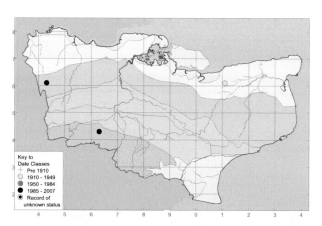

A rare and little known species in the county. The life cycle is likely to be very similar to the closely related *P. perturbator* and *P. susterai*. The species is found in other counties from various biotopes. Prey is not known but is most likely to be spiders. One of the Kent records cannot be mapped as the locality information is too vague. The adult wasps will visit *Euphorbia* for nectar.

Occurrence: 2, 3 iv-vi 1936-2006
National status Falk: pNa
Kent status Present work: pKRDB2

Priocnemis exaltata (Fabricius)

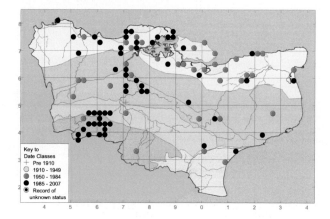

Although recorded mainly in the north of the county, in Maidstone and Tunbridge Wells, this species is probably much more widely distributed than this. One of the larger and more easily identified species of the genus. Most records are from the sands and London clay. Prey includes wandering spiders of the families Lycosidae, Salticidae and Pisauridae. It is believed the species stocks multiple cells at the end of a single burrow.

Occurrence: 55, 93   *vi-x*   1897-2005
No status.

Priocnemis fennica Haupt
Once confused under the name "*P. femoralis*", which included *P. hyalinata*.

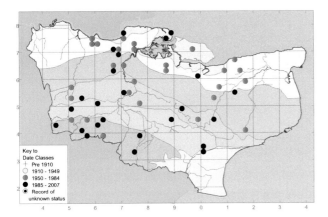

A widely distributed and frequent species but apparently scarce on the Weald Clay. Often found in open, broad-leaved woodland such as chestnut coppice. Prey includes Lycosidae but reliable records are scarce, as the species was only recognised as British in 1979.

Occurrence: 24, 52   *vi-ix*   1980-2007
No status.

Priocnemis gracilis Haupt

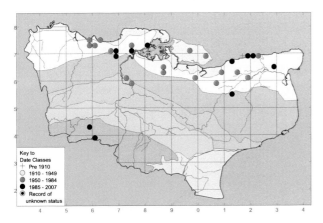

A scarce species but probably under recorded. The species is found mainly in the north of the county with only a few records in the south. Found predominantly in woodland and coastal marsh in Kent. Prey includes Salticidae and Clubionidae.

Occurrence: 11, 30   *vii-x*   1896-2002
National status   Shirt: RDB3   Falk: pNb
Kent status   Waite: Notable   Present work: pKb.

Priocnemis hyalinata (Fabricius)
Once confused under the name "*P. femoralis*", which included *P. fennica*.

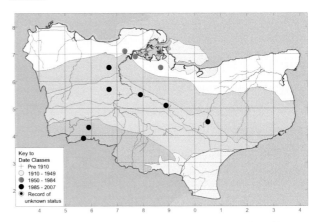

A very scarce species occurring mainly on the sands, with some data from chalk scarps, Prey includes Lycosidae. There are no known flower visit data.

Occurrence: 7, 11   *vi-ix*   1896-2002
National status   Falk: pNb
Kent status   Waite: Notable   Present work: pKa

Priocnemis parvula Dahlbom

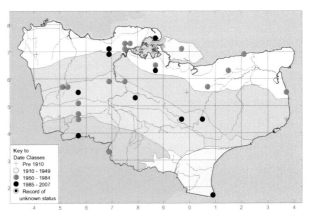

Not a common species in the county but possibly under recorded. Occurs on most soil types but scarce on the chalk, where it may be present only on the scarps. Occasionally coastal. In other counties is strongly associated with sandy heath. Prey are particularly Lycosidae but also Thomisidae and Salticidae.

Occurrence: 10, 30   *vi-ix*   1890s-2007
Kent status   Present work: pKb

Priocnemis perturbator (Harris)
Was once included with *P. susterai* under the name "*P. fuscus*".

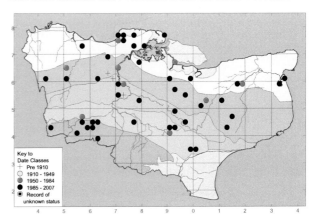

The species overwinters as adults and emerges in the spring. Males tend to be scarce. One of our larger pompilids and fairly common. Occurs on most soil types, including frequently on the chalk, but is rarely coastal. Is said to prefer dryer soils. Due to the confusion with *P. susterai*, there are no reliable prey records.

Occurrence: 41, 53 iv-vi 1894-2005
No status.

Priocnemis pusilla Schioedte

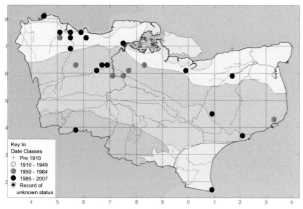

Probably very under recorded in the county. Absent from clay soils and not recently recorded from the Lower Greensand. Present on the chalk and occasionally coastal. It is said usually to occur on lighter soils. Reliable prey records include Clubionidae and Salticidae.

Occurrence: 17, 25 vi-x 1980-2001
No status.

Priocnemis susterai Haupt

Synonymy: *P. clementi*. Was once confused with *P. perturbator* under the name "*P. fuscus*".

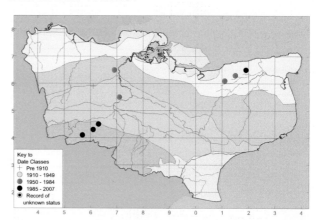

A very scarce species in Kent, with a life cycle similar to the more common *P. perturbator*. The distribution is strongly and positively correlated in the county with major woodland. There are no reliable prey records.

Occurrence: 4, 9 v-vi 1899-2004
Kent status Present work: pKa

Subfamily Pompilinae

Most of the species considered in this subfamily were once placed in a single blanket genus, *Pompilus*. This has been partitioned into defendable units, many of which are now regarded as genera.

The species of *Arachnospila* can prove difficult to identify, particularly the females of the *A. anceps* group. The genus *Evagetes* contains cleptoparasitic forms or those suspected of being so. The genus is not always readily recognisable.

Evagetes dubius and *Anoplius viaticus* have apparently been recorded from Kent (Day, 1988) but no data for these species have been located.

Pompilus cinereus (Fabricius)
Synonymy: *P. plumbeus*.

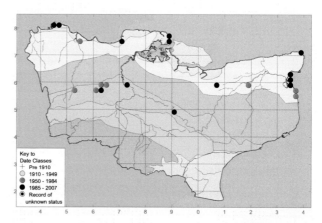

Probably very under recorded – for instance the species is likely to occur at Dungeness. Most modern localities are coastal but there are a few on the Lower Greensand and one on the Eocene sands. Has a preference for open sand. The prey are wandering spiders, particularly Lycosidae, but other families are also taken.

Occurrence: 13, 21 vi-ix 1850s-2005
Kent status Present work: pKb

Priocnemis schioedtei Haupt

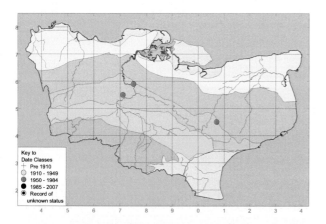

A very rare species in the county but known to have occurred at Oaken Wood, Barming, over a period of nearly a century. However, there are no modern records for the county. A species of openings in broad leaved woodland. Two of the three Kent localities are typical but one is from chalk scarp. Prey includes Clubionidae.

Occurrence: 0, 3 vii-viii 1896-1982
National status Falk: pNb
Kent status Waite: Notable Present work: pKRDB1

Agenioideus cinctellus (Spinola)

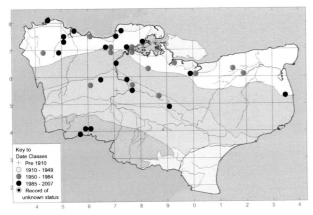

I hesitate to give this species scarcity status, as it is small and agile, hence probably very under recorded. The distribution of data through time is also unusual. Particularly frequent on sandy soils and sometimes estuarine, but not strictly coastal. This is a species frequent on old walls and may once have been associated with sandstone cliffs. It also nests in a wide variety of natural cavities and abandoned bee and wasp cells. Prey is principally Salticidae and occasionally Thomisidae.

Occurrence: 20, 35   *v-ix*   1896-2007
No status.

Agenioideus sericeus (Vander Linden)

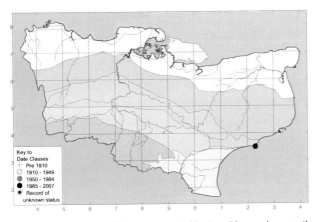

This species is reported here as new to Kent and has only recently been found in mainland Britain, although known from the Channel Islands. A female was captured at Folkestone, hunting near a horizontal spider web. Abroad, the species is reported as nesting in sand and loose stone walls, preying on Salticidae, Pisauridae and Thomisidae.

Occurrence: 1, 1   *vii*   2006
Kent status   Present work: pKRDBK

Arachnospila anceps (Wesmael)
In older literature misidentified as *"Pompilus trivialis"*.

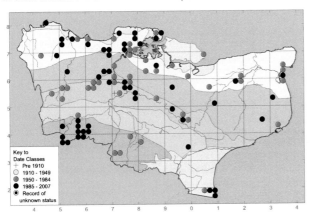

Females of some *Arachnospila* are particularly difficult to identify. *A. trivialis* is now regarded as a species distinct from *A. anceps*. The present species is common and widely distributed, mostly on the sands but also on chalk and clays. One of the county's most frequently recorded pompilids and sometimes captured in gardens. A wide range of prey is captured, including Lycosidae, Clubionidae and Thomisidae but also other families. The prey is hidden on a plant while the burrow is dug.

Occurrence: 52, 98   *iv-x*   1896-2007
No status.

Arachnospila consobrina (Dahlbom)

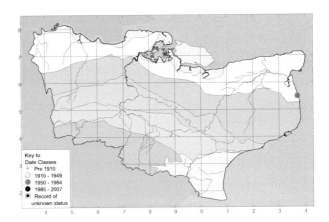

This species appears exceptionally rare in the county but it is plausible this is due in part to misidentification - it is a difficult species to identify correctly. An old record from Oaken Wood, Barming, is thought to be in error: specimens in the Maidstone Museum collection appear to be *A. anceps*. Nationally, very much a coastal dune species. So far only recorded with certainty from Deal in the county. There are few reliable prey records but may include Segestriiidae.

Occurrence: 0, 1   *No Kent flight period can be given*   1980s
National status   Shirt: RDB3   Falk: pRDB3
Kent status   Present work: pKRDB1

Arachnospila minutula (Dalhbom)

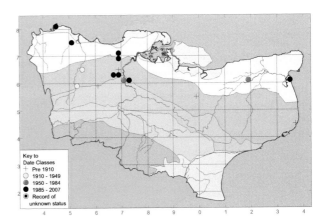

A scarce species but well known from the county. Recorded particularly from the chalk, but is sometimes coastal and can be found on sandy soils. The prey is at least in part Lycosidae. The spider is stung and an egg laid. The spider is left *in situ*, recovers and is eventually killed by the growing wasp larva, parasitoid-like.

Occurrence: 8, 16   *vi-ix*   1890s-1998
National status   Falk: pNb
Kent status   Waite: Notable   Present work: pKa

Arachnospila spissa (Schioedte)

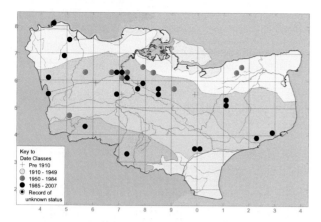

Frequent on the chalk but scarcer elsewhere. Often in woodland, occasionally estuarine but rarely coastal. Probably has a similar biology to *A. minutula*. Prey are Lycosidae.

Occurrence: 21, 36 *v-viii* 1896-2006
No status.

Arachnospila trivialis (Dahlbom)

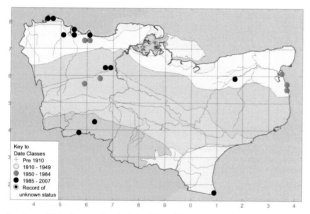

Another difficult species to identify. Although widely distributed, it does not appear to be common. Frequently a species of coastal dunes and other open, sandy soils. Prey include Thomisidae and possibly Lycosidae.

Occurrence: 12, 19 *v-ix* 1978-2001
Kent status Present work: pKb

Arachnospila wesmaeli (Thomson)

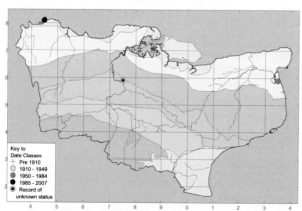

Another difficult species to identify in the female. Rare in the county although possibly under recorded. Only firmly known from estuarine and coastal sands, the inland record being uncertain. Prey are mainly Lycosidae.

Occurrence: 1, 4 *v-viii* 1872-1997
National status Shirt: RDB3 Falk: pNa
Kent status Waite: Notable Present work: pKRDB2

Evagetes crassicornis (Shuckard)

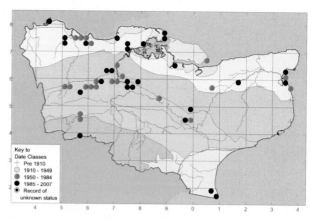

The most common British cleptoparasitic pompilid. Frequent on sandy soils and the chalk scarps. The hosts are particularly likely to include *Arachnospila anceps* and possibly *A. trivialis* but field observations are scarce.

Occurrence: 27, 51 *v-ix* 1890s-2001
No status.

Evagetes pectinipes (Linnaeus)

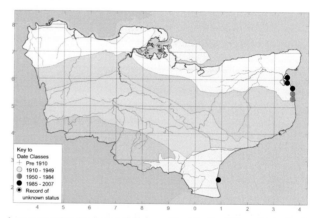

A rare species known mainly from Kent in the UK and principally from the Deal to Sandwich dunes. The species is well recorded from its main range but it remains unclear if the Romney Sands specimen was merely a vagrant. A cleptoparasite, most probably of *Episyron rufipes*, a mainly coastal species.

Occurrence: 4, 6 *vi-ix* 1966-2003
National status Shirt: RDB1 Falk: pRDB1
Kent status Waite: KRDB3 Present work: pKRDB3

Anoplius concinnus (Dahlbom)

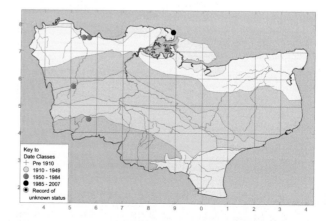

Anoplius species are frequently associated with wet habitats. Although the present species has no national scarcity status, it is of rare occurrence in the county. Only known from Vice County 16 by one recorder over an eight-year period. It is not generally well known from south-east England. Said to be a species of stony or gravely places, including stream sides. Prey are particularly Lycosidae, stocked in multicellular nests.

Occurrence: 1, 5   *vi-ix*   1980-1987
Kent status   Present work: pKRDB3

Anoplius infuscatus (Vander Linden)

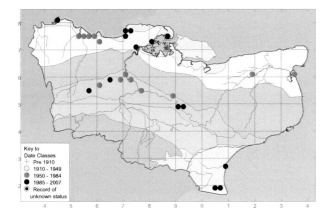

Found widely on moist, sandy soils but not everywhere. May be absent from the High Weald sands. Frequently estuarine and coastal. Prey include mostly Lycosidae but also Agelenidae and Thomisidae. The nest is a short burrow in the sand and the prey hung on low vegetation whilst it is constructed.

Occurrence: 14, 27   *vi-ix*   1961-2003
No status.

Anoplius nigerrimus (Scopoli)
Once confused with other species under the old name "*Pompilus niger*".

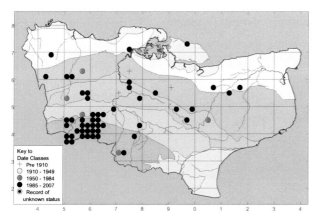

The *Anoplius* species least associated with wet habitats. Mostly from sandy soils but not usually estuarine or coastal. Very common in the Tunbridge Wells area but less frequent elsewhere. Nests under stones, in hollow plant stems and snail shells etc. Reported prey are Lycosidae, Gnaphosidae and Pisauridae.

Occurrence: 50, 58   *v-x*   1896-2007
No status.

Episyron rufipes (Linnaeus)

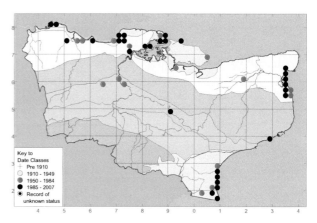

Almost entirely estuarine and coastal, but has been found inland near Maidstone in sand pits. Particularly a species of coastal dunes in Kent. Both sexes can be found on flowers, particularly Apiaceae. Prey are mainly Araneidae and occasionally Lycosidae. The spider is hung on low vegetation while the burrow is dug in loose sand.

Occurrence: 29, 45   *v-x*   1850s-2004
No status.

Aporus unicolor Spinola

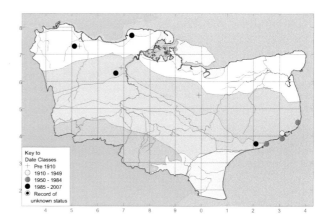

This has never been a frequent species in the county although known for a long time. Almost entirely confined to chalk soils in Kent, even where coastal. Often taken on wild carrot, *Daucus carota* (Apiaceae). A specialist on the purse-web spider, *Atypus affinis*, which is paralysed in its own burrow. Development of the *Aporus* larva takes place there.

Occurrence: 4, 11   *vii-viii*   1850s-2000
National status Falk: pNa
Kent status Waite: Notable   Present work: pKRDB3

Subfamily Ceropalinae

There is only one British genus in this subfamily, the type, *Ceropales*. The ceropalines are exclusively cleptoparasitic and only one of the two British species has been found in Kent.

Ceropales maculata (Fabricius)

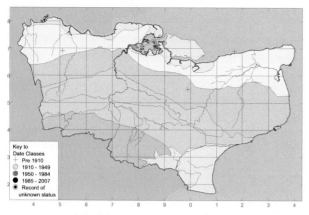

Only known from three very old records in the county. Assumed extinct here. Was sometimes coastal. Both sexes visit Apiaceae. A cleptoparasite of various pompiline species and occasionally of *Priocnemis exaltata* (Pepsinae).

Occurrence: 0, 3   *viii*   1850s-1890s
Kent status   Present work: pKRDB1+

Family Vespidae

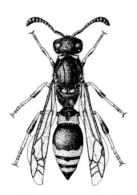

♀ *Symmorphus gracilis*
with wings pleated closed at rest

The pleated-winged or true wasps are represented in Britain by two subfamilies, Eumeninae and Vespinae. *Polistes dominulus* is now included in the subfamily Vespinae, tribe Polistini.

The British eumenines are all solitary species, each female provisioning her own nest, whilst vespines are social wasps, the Vespini having morphologically defined queens. Polistini holds more primitive social wasps with smaller colonies than Vespini.

Subfamily Eumeninae

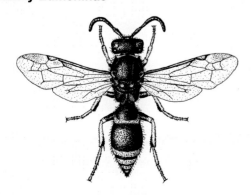

♀ *Ancistrocerus gazella*

The 16 Kentish species are mason wasps. I can find no data for the heath potter wasp, *Eumenes coarctatus*, in Kent. Eumeninae construct nests from mud (*Ancistrocerus oviventris*), or more usually, of mud partitions in cavities. A few species, in *Odynerus*, burrow in the ground, raising "chimneys" above the nest hole. Eumenine prey are invariably larval insects, mostly of the orders Lepidoptera and Coleoptera but rarely of symphytan Hymenoptera. These are usually hunted on vegetation. The egg is laid in the empty cell, suspended on a thread secreted by the female wasp, before provisioning.

The adult wasps will visit flowers for nectar and take honeydew secreted by Hemiptera from leaves. The Kent species were once placed in a very broad genus *Odynerus* but this name is now restricted to four British (and only two Kent) species.

Parasites particularly include rubytailed wasps such as the *Chrysis ignita* complex, which oviposit in the eumenine cells. The *Chrysis* larva devours the mature wasp larva or pre-pupa.

Odynerus melanocephalus (Gmelin)

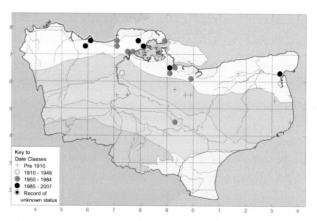

Modern records are from the London clay and Eocene sands. The species is in clear decline. The female nests in clay soil, in flat, sloping or vertical faces. Prey are weevil- and small lepidopterous larvae. A host of the chrysidid wasp *Chrysis viridula*.

Occurrence: 6, 26   *v-vii*   1850s-2005
National status   Falk: Na
Kent status   Waite: Notable   Present work: pKRDB3

Odynerus spinipes (Linnaeus)

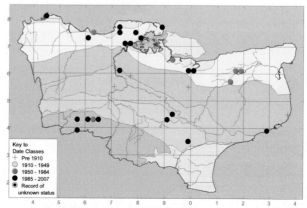

Not a common species in Kent but neither is there strong evidence for a decline. Mainly on sandy and clay soils, including scarp faces. Generally nests in vertical banks, both of clay and sand. Prey are *Hypera* larvae (Coleoptera). Host of the chrysidid wasps *Chrysis viridula* and *Pseudospinolia neglecta*.

Occurrence: 20, 30   *v-vii*   1897-2007
No status.

Gymnomerus laevipes (Shuckard)

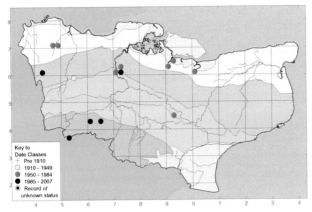

This is a rare species in the county and possibly declining. It may be under recorded here. On chalk and the sands but, apparently, not recorded from clay soils. Nests in hollow plant stems and preys on weevil larvae of the genus *Hypera*. The adults take nectar from a variety of flowers with shallow corollas. There are no known chrysidid parasitoids.

Occurrence: 5, 14   *v-viii*   1947-2005
Kent status   Present work: pKRDB3

Microdynerus exilis (Herrich-Schaeffer)

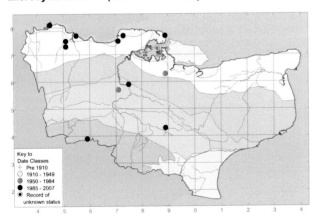

A species known to be extending its range but not particularly frequent in Kent yet. Found on most soil types. The female nests in tubular cavities such as beetle exit holes in fence posts. Prey are weevil larvae.

Occurrence: 10, 12   *vi-viii*   1951-2001
National status   Falk: pNb
Kent status   Waite: Notable   Present Work: pKa

Ancistrocerus antilope (Panzer)

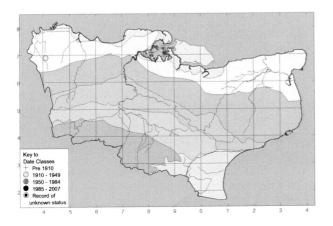

Ceropalinae to Vespidae; Eumeninae

Believed extinct in the county and, nationally, there has been a recent, significant decline. Nests in stems of bramble and elder and the adults take nectar from various flowers with shallow corollas. Prey are usually lepidopterous larvae but occasionally those of beetles and sawflies. Host to the rare *Chrysis pseudobrevitarsis* (not Kentish) and *C. longula*.

Occurrence: 0, 2   *v-vi*   1893, 1925
National status   Shirt: RDB3   Falk: pRDB3
Kent status   Present work: pKRDB1+

Ancistrocerus gazella (Panzer)

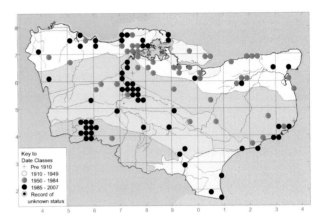

Common and widely distributed. Most frequent on sandy soils but also on coastal chalk, occasionally from chalk scarps. Nests in tubular cavities, such as beetle exit holes in wooden posts. Cells are defined by mud partitions. The prey are lepidopterous caterpillars. A probable host to *Chrysis angustula*, *C. ignita* and *C. impressa*.

Occurrence: 61, 110   *v-ix*   1897-2007
No status.

Ancistrocerus nigricornis (Curtis)
Synonymy: *A. callosus*

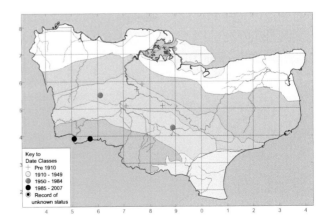

A very rare species in Kent, formerly a little more common but known to be nationally declining. Old records are mostly from the sands but those of 1983 from weald clay. Prey and nest as previous but often in bramble and elder stems. The prey are small lepidopterous larvae. There are no known chrysidid parasitoids.

Occurrence: 2, 8   *v-ix*   1895-2006
Kent Status   Present Work: pKRDB2

Ancistrocerus oviventris (Wesmael)
Synonymy: *A. pictus*

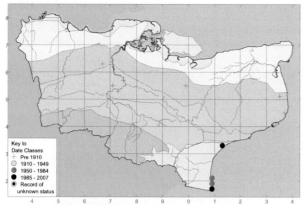

Scarce and declining. Very old records from sands and chalk inland, but all modern and fairly modern data are coastal. A mud dauber, usually nesting in crevices such as those in walls. Prey are small lepidopterous larvae. Host to *Chrysis ruddii* and possibly also *C. rutiliventris*.

Occurrence: 2, 11   v-vii   1896-2006
Kent status   Present work: pKRDB2

Ancistrocerus parietinus (Linnaeus)

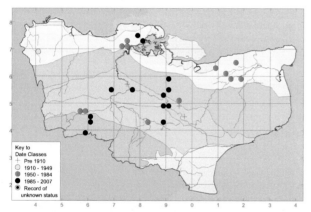

Found on all the main soil types in the county. I hesitate to give this species county scarcity status, given the widespread national distribution. Usually nests in hollow stems such as those of brambles but the name suggests, perhaps erroneously, that it will nest in crevices in walls. Prey are lepidopterous larvae. A putative host of *Chrysis longula*.

Occurrence: 13, 29   v-viii   1894-2006
No status.

Ancistrocerus parietum (Linnaeus)

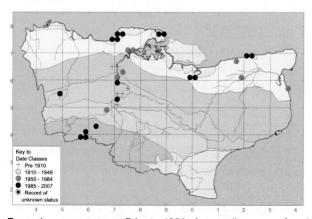

Formerly more common. Prior to 1954, *A. gazella* was confused under the name *A. parietum* with the present species. Found on all the main soil types in the county, especially in the Medway valley. Prey and nest much as in *A. gazella*. There may be a loose association between this species and water bodies, it visits *Scrophularia auricularia*. Host to *Chrysis ignita* and probably *C. impressa*.

Occurrence: 17, 31   v-ix   1896-2007
No status.

Ancistrocerus scoticus (Curtis)
Synonymy: *A. albotricinctus*

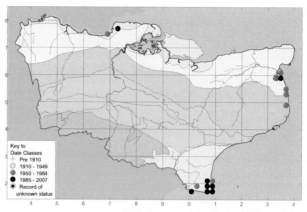

Declining nationally, showing a retreat to coastal sand and shingle. In Kent has always for the most part been coastal. Nests in exposed crevices, preying on lepidopterous and weevil larvae. A probable host of *Chrysis rutiliventris*.

Occurrence: 8, 18   v-ix   1896-2004
Kent status   Present work: pKb

Ancistrocerus trifasciatus (Mueller)
Synonymy: *A. trimarginatus*.

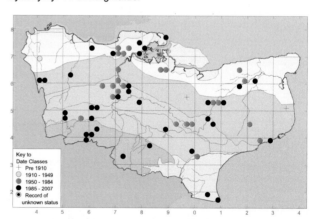

Found widely across the different soil types in the county. Only rarely coastal. The species usually nests in dead plant stems and the prey is lepidopterous and weevil larvae. Probably the main host of *Chrysis angustula*.

Occurrence: 35, 62   v-ix   1896-2007
No status.

Symmorphus bifasciatus (Linnaeus)
Synonymy: *S. sinuatissimus, S. mutinensis*.

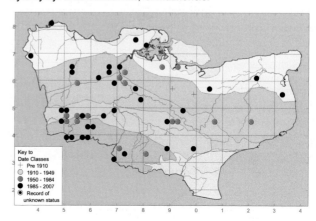

Often found in damp habitats but only rarely on clay soils. Less common than *S. gracilis* but widely distributed across the county. May be a little under recorded. Tube nester, often in dead wood, and crevices in old walls. The prey is the larva of *Phyllodecta vulgatissima* (Coleoptera). There are no known chrysidid parasitoids in Britain.

Occurrence: 34, 53   vi-ix   1890's-2007
No status.

Symmorphus connexus (Curtis)
In older literature misidentified as *S. bifasciatus*.

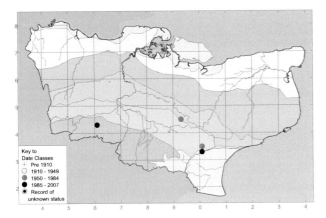

A very rare species in the county whose limited distribution is correlated with open, broad leaved woodland. It is undergoing a decline nationally. Nests in dead wood, hollow twigs and perhaps old walls; prey are lepidopterous- and weevil larvae. Adults visit shallow flowers for nectar. There are no known chrysidid parasitoids for this species in Britain.

Occurrence: 2, 4   vi-vii   1967-2001
National status   Shirt: RDB3   Falk: pRDB3
Kent status   Waite: KRDB1   Present work: pKRDB1

Symmorphus crassicornis (Panzer)

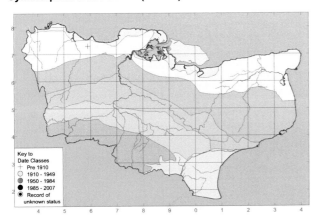

Only known from two 19th century Kent records (the same locality) and probably extinct in the county. However, the species has recently been refound at several localities in Surrey and so may be overlooked in Kent. Declining nationally. Nests in dead wood and hollow plant stems, preying on the larva of the leaf beetle, *Chrysomela populi*. Recently shown to be the host of *Chrysis fulgida*.

Occurrence: 0, 1   vii   1837, 1896
National status   Shirt: RDB3   Falk: pRDB3
Kent status   Present work: pKRDB1+

Symmorphus gracilis (Brulle)
Synonymy: *S. elegans*.

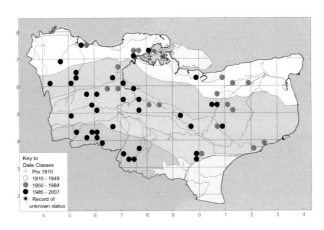

The most common Kentish *Symmorphus*, occurring on all of the main soil types. Usually nests in dead wood and stems near damp areas. Loosely associated with Common figwort. The female wasp frequently preys on the larva of the weevil, *Cionus hortulanus*, and both sexes of the *Symmorphus* will visit the figwort flowers for nectar. Males have also been taken on wood spurge. No known chrysidid parasitoids.

Occurrence: 38, 59   v-viii   1895-2006
No status.

Subfamily Vespinae

The social wasps, with one indigenous tribe Vespini and one vagrant species of Polistini.

Tribe Polistini. A species of *Polistes* is occasionally found in Kent. This is *P. dominulus*, not a native British wasp and so far not established in the county.

Polistes dominulus (Christ)

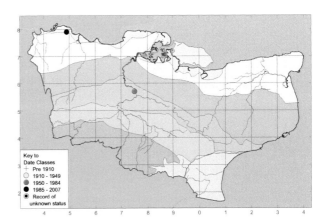

This is a social paper wasp with fairly small colonies. At present it is classified as a vagrant in the county. Recent observations in Surrey suggest that the species may establish (D W Baldock, *pers. comm.*). The nest does not have a paper envelope and consequently is susceptible to rain. The Erith specimen was a male and may have come from a successful nest of that year, but the species has not been refound there more recently.

Occurrence: 1, 2   i, ix   1958, 1992
Kent status   Present work: Not native.

Tribe Vespini.

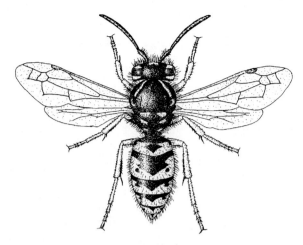

Worker *Vespula vulgaris*

There are eight Kent Vespini. These are the social species which everyone associates with the word "wasp". Two species, *Dolichovespula media* and *D. saxonica*, are relatively recent colonists, the former being first found in the county in 1986 and the latter in 1994. Both of these species are frequent on the near continent and have become so in Kent. The hornet, *Vespa crabro*, has dramatically increased in numbers and range in Kent since 1995, as it has in other southern counties. Some Vespini declined in numbers in the early 1980s (Dr M E Archer, *pers. comm.*), thought to be related to climatic conditions, including *Dolichovespula norwegica*, *Vespula germanica* and *V. rufa*. Flight periods are rather meaningless in Vespini, as queens can be found in flight on very mild, sunny days throughout the winter. Social wasps are not a popular group with recorders and so are under recorded.

Vespa crabro Linnaeus

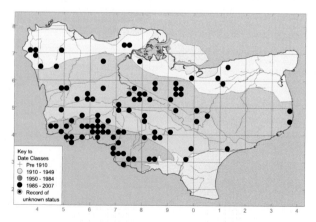

This species, the hornet, apparently goes through cyclical periods of abundance possibly related to climate changes, but has never been as frequently recorded as at present in Kent. It is sometimes found in large numbers, although colony size is compatively small. There is a strong south westerly bias to the data, possibly indicating a spread from Surrey and East Sussex. The hornet occurs widely across the soil types and is mainly a woodland species. Nests tend to be constructed in standing hollow trees. They are sometimes initiated in bird nesting boxes by the queen and translocate to a larger cavity when there are several workers present.

Occurrence: 92, 94   *ii-xi*   1891-2007
No status.

Dark ♂ *Dolichovespula media* on flowerhead of *Echinops*
© Amanda Brookman

Dolichovespula media (Retzius)

A species now well known in Kent. It is very under recorded and may be generally distributed across the county. The queens resemble small worker hornets whilst the workers look more like typical social wasps but usually larger and darker. Does not appear to be ecologically restricted. The nest is frequently constructed in hedges, sometimes resulting in conflict with gardeners, but also nests in fruit trees and many other aerial situations.

Occurrence: 47, 47   *iii-xi*   1986-2007
National status Shirt: RDB3   Falk: pNa
No Kent status.

Dolichovespula norwegica (Fabricius)

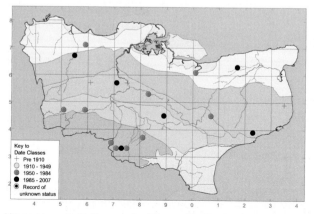

Never common in the county, this species is clearly declining. In some other southern counties there is evidence that it is being replaced by *D. saxonica*. Nests are built in similar sorts of situations to *D. media* and contain a few hundred workers at most, when mature, similar to other *Dolichovespula*. As with all social wasps, adults visit Apiaceae and other flowers for nectar.

Occurrence: 6, 20 *iv-ix* 1888-2006
Kent status Present work: pKa

Dolichovespula saxonica (Fabricius)

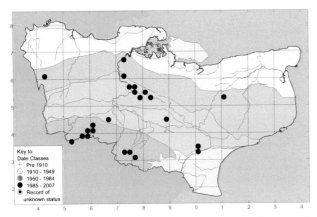

First recorded from Britain in 1987 and Kent in 1994. However, the species has spread rapidly, and is certainly under recorded. There is some evidence that it is replacing *D. norwegica*, although this latter wasp has never been common in Kent. Nests tend to be constructed more often in hollow trees than is typical for *D. norwegica*.

Occurrence: 23, 23 *iii-viii* 1994-2006
National status Falk: pRDBK
No Kent status.

Dolichovespula sylvestris (Scopoli)

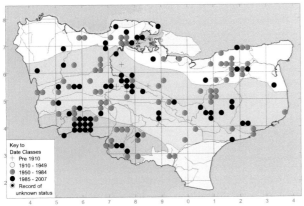

The most common Kent *Dolichovespula*, showing great adaptability in choice of nesting site. Species of this genus generally nest above ground but *D. sylvestris* will exploit cavities below ground as well. It will also nest in empty bee hives and under eaves. Found fairly generally across the county, although possibly absent from the metropolitan area and rare in salt marsh habitats. Does not do well on clay soils.

Occurrence: 64, 147 *iv-ix* 1897-2006
No status.

Vespula germanica (Fabricius)

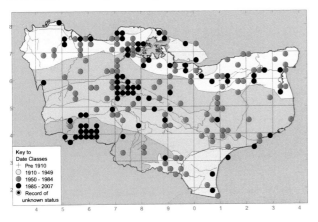

One of the two species regarded by the layperson as "the Wasp". Widespread across the county but apparently scarcer on the chalk. Predominantly an underground nester and will vindictively defend its nest against vertebrate predators (including people), although some nests fall prey to badgers. A less synanthropic species than *V. vulgaris*. Mature nests contain a few thousand workers, which can be found from mid May to November or December. Nests occasionally over winter but do not continue into the new season in the UK.

Occurrence: 78, 193 *i-xii* 1887-2007
No status.

Vespula rufa (Linnaeus)

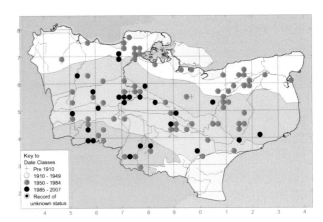

This species has never been particularly frequent in the county but after showing a decline in the early 1980s is possibly recovering. Found on various soil types including the chalk. Predominantly an underground nester, although the nests are sometimes close to the surface. The workers number only a few hundred in the mature nest and tend to be docile. Host of the workerless social parasite, *Vespula austriaca*, which has, however, never been recorded from the county.

Occurrence: 25, 96 *iv-ix* 1894-2007
No status.

Vespula vulgaris (Linnaeus)

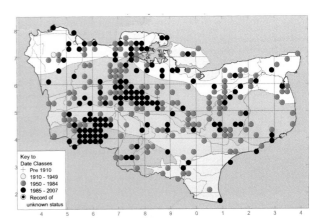

The second of the species considered to be "the Wasp". An abundant species most frequently recorded from sandy soils. Much more synanthropic than *V. germanica*. Workers appear in the predominantly underground nests in mid June and at the peak can number a few thousand. Will on occasion nest in roof spaces of houses, possibly surviving here well into winter or even to the spring. However, as with *V. germanica*, these nests in the UK do not continue into the new season. This is a defensive species that stings readily when in conflict with people.

Occurrence: 125, 284 *i-xii* 1887-2007
No status.

Superfamily Apoidea

In this volume the "family Sphecidae" of Richards (1980) is partitioned into two British families, Sphecidae (*sensu stricto*) and Crabronidae, and placed in the superfamily, Apoidea. The subfamilies recognised by Richards (1980) are to be found within these two, but some are downgraded to tribal status. Apoidea also includes the bees (family Apidae). Two more families related to the former "sphecoids" occur abroad. The cladistic analyses of Melo (1999) suggest that the bees are closer to the Crabronidae (sometimes arising within the latter and hence rendering that family paraphyletic) than either is to the Sphecidae (*sens. str.*). Logically, therefore, the bees are ranked as only one family.

Family Sphecidae (*sens. str.*)

The Sphecidae in the strict sense contains only two British genera, *Podalonia* and *Ammophila*, with a further, *Sphex*, in the Channel Isles. The mainland British genera fall in the subfamily Sphecinae, tribe Ammophilini.

Subfamily Sphecinae

Tribe Ammophilini.

♀ *Ammophila sabulosa* © Jeremy Early

These are known as sand wasps in Britain, digging unicellular nests in sandy soils. There are four British species, two in each genus and three have been firmly recorded from the county. *Ammophila pubescens*, the heath sand wasp, has occasionally been claimed from Kent but I have not been able to confirm any specimens (but see Appendix 1).

Ammophila sabulosa (Linnaeus)

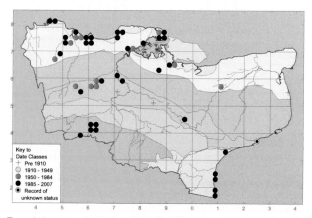

Found almost exclusively on sandy soils and can be locally frequent in these situations. The species is not easily overlooked and the dot map is probably a fair representation of the Kent distribution. However, the absence of modern records from the Deal/Sandwich area may be an artifact of recording. Prey are lepidopterous or more rarely, symphytan larvae. The nest is a short burrow in sandy soil with only one cell. Each cell is provisioned with only one caterpillar, with rare exceptions.

Occurrence   35, 51   *v-x*   1850s-2006
No status.

Podalonia affinis (Kirby)

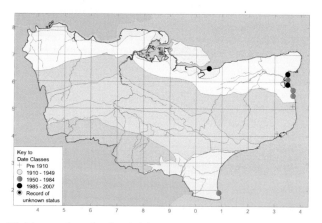

Well known from the Sandwich dunes and recently recorded from Graveney Marshes (coll. L Clemons). There are older records from Deal, Dover and Dungeness. The females pass the winter as adults and can appear on the wing as early as March. Prey are lepidopterous caterpillars, captured before the nest is dug.

Occurrence   3, 9   *iii-ix*   1850s-2005
National status   Shirt: RDB3   Falk: pRDB3
Kent status   Waite: KRDB2   Present work: pKRDB2

Podalonia hirsuta (Scopoli)

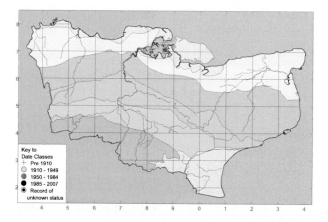

Believed to be extinct in the county, if indeed of native Kent occurrence. The single record may be of a vagrant specimen as the determination (E. Saunders) is certainly reliable.

Occurrence   0, 1   *A Kent flight period cannot be given*   1890
National status   Falk: pNb
Kent status   Present work pKRDB1+.

Family Crabronidae

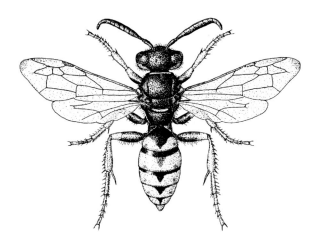

♀ *Philanthus triangulum*

Known by vernacular names such as "digger wasps" and "black wasps", the Crabronidae form the second largest family of British aculeates, surpassed in size only by the bees, family Apidae. Crabronidae exhibit great diversity, as could be expected by the large number of species. Many of the larger forms are boldly marked with yellow but most are small, largely black and go completely unnoticed by the layperson. The classification used here follows Melo (1999) in recognising only five subfamilies.

Subfamily Astatinae

There are two British genera in this subfamily, each with one indigenous species. Both are found in Kent.

Astata boops (Schrank)

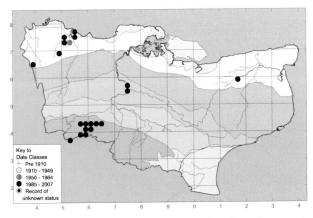

A rather scarce wasp in the county but found on the various sands. Nationally, it shows a distinctly restricted, southern distribution. The female nests in sandy soil and preys on the nymphs of heteropterous bugs, particularly of the family Pentatomidae. The adult wasps will visit Apiaceae for nectar. Host of *Hedychridium roseum* (Chrysididae).

Occurrence: 19, 20   *vi-x*   1979-2007
No status.

Dryudella pinguis (Dahlbom)

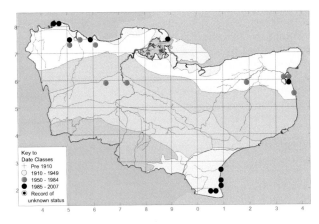

This species has a wider national distribution than *Astata boops* but is recorded less frequently. Again it shows a preference for sandy soils, with most modern data in the county coastal. The female preys on Heteroptera, of the families Pentatomidae and Lygaeidae. It is a host of the chrysidid wasp *Hedychridium cupreum*.

Occurrence: 11, 20   *vi-ix*   1856-2003
Kent status Present work: pKb

Subfamily Crabroninae (including the former Larrinae)

Crabroninae, as defined here, is a large, diverse group. The subfamily is divisible into six British tribes: Mellinini (transferred from the Bembicinae), Larrini, Miscophini, Trypoxylini, Crabronini and Oxybelini. Miscophini and Trypoxylini might in fact be very close to, or synonymous with, Crabronini but are retained here.

Tribe Mellinini. A borderline candidate for subfamily rank, although perhaps best treated as a basal branch or clade of the Crabroninae. One of the two British species, *Mellinus crabroneus*, is probably extinct in the UK and not known from Kent. The other, *M. arvensis*, is frequent in the county. The species are predators of Diptera.

Mellinus arvensis (Linnaeus)

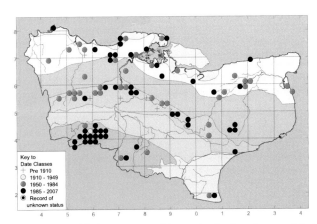

Although classified as common, this species is strongly associated with sandy soils; known from sedimentary, alluvial and coastal sands. The mated female burrows in sandy soil to construct the nest and provisions with paralysed flies. Particularly preyed on are the families Muscidae (*sens. lat.*), Stratiomyidae and Syrphidae. The species is parasitised by miltogrammine flies, the larvae of which feed on the prey provided by the host female for her own larvae.

Occurrence: 49, 94   *vii-xi*   1894-2007
No status.

38 The Bees, Wasps and Ants of Kent

Tribe Larrini. There is only one British genus, *Tachysphex*, the species of which are orthopteroid predators. There are three recorded Kent species.

Tachysphex nitidus Vander Linden
Once confused under the name *T. unicolor*, from which it was separated recently. *T. unicolor* is also a British species.

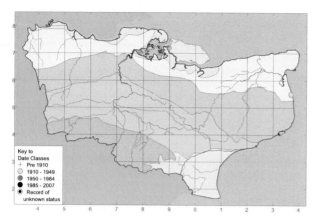

A rare species in the county, with only two confirmed records (coll. F Smith & K M Guichard, det J C Felton). It nests in sandy soil and preys on grasshopper nymphs of the family Acrididae (Orthoptera).

Occurrence: 0, 1   v   1850s, 1939
Kent status   Present work: pKRDB1+

Tachysphex obscuripennis (Schenck)

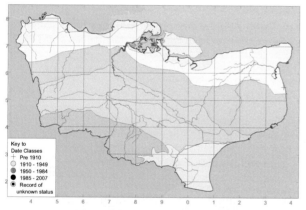

Only known from one British record, of a few males captured on the Deal sandhills in the nineteenth century and with no evidence of breeding (coll. E Saunders). Hence it is here not thought to be a British species. Abroad, it preys on *Ectobius* nymphs (Blattodea).

Occurrence: 0, 1   viii   1882
National status   Shirt: Appendix   Falk: Appendix
Kent status   Present work: Probably not native.

Tachysphex pompiliformis (Panzer)

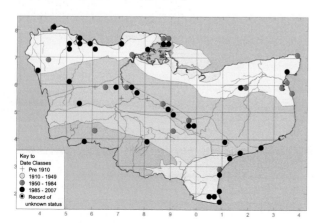

A common species usually found on light, sandy soils; at Folkestone Warren is present on blown shell sand. The prey are acridid nymphs. A host of the chrysidid wasps *Chrysis illigeri*, *Hedychridium ardens* and possibly *H. roseum*.

Occurrence: 34, 53   v-ix   1895-2006
No status.

Tribe Miscophini.

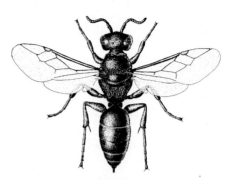

♀ *Nitela lucens*

This tribe has two British genera, *Nitela* first being recorded from Kent in 1982 (Allen, 1983). The taxonomy of the European *Nitela* was put on a sound basis by Gayubo and Felton (2000). Both *Nitela* and *Miscophus* have two Kent species.

Miscophus ater Lepeletier
Synonymy: *M. maritimus*.

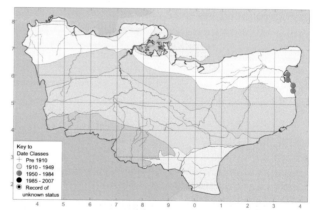

A very rare wasp in Britain, known only from the Deal/Sandwich area in Kent and from Camber, East Sussex. It occurs only on the dunes. It is important to confirm the continued presence of the species in Kent. The recorded prey is entirely small spiders.

Occurrence: 0, 4   vi-ix   1850s-1984
National status   Shirt: RDB2   Falk: pRDB2
Kent status   Waite: KRDB1 Present work: pKRDB1

Miscophus concolor Dahlbom

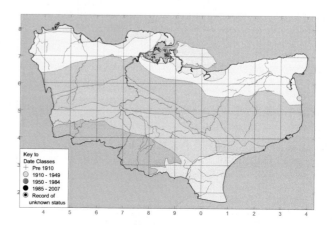

This is a species of dry sandy heath and the record from Deal is very unusual (coll. K M Guichard). The sole Kent specimen is a male, hence difficult, and may be a misidentification for *M. ater*. The prey of *M. concolor* is recorded as small spiders.

Occurrence: 0, 1   *v*   1939
Kent status   Present work: pKRDBK

Nitela borealis Valkeila

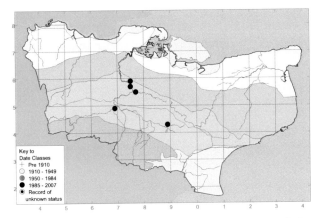

First recorded as British from Kent in 1982 and there remain few British records outside the county. The female nests mainly in dead wood, often in wooden posts containing beetle exit holes. There are occasional records of nests in holes in mortar and brickwork. The only recorded prey in Britain are psocid nymphs.

Occurrence: 5, 5   *v-viii*   1982-1994
National status   Shirt: RDB3   Falk: pRDBK
Kent status   Waite: KRDB1   Present work: pKRDB2

Nitela lucens Gayubo & Felton
Synonymy: *N*. sp. nr. *spinolae*.

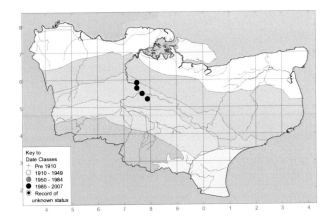

First recorded as British from Surrey in 1982 and is more widely distributed in southern Britain than *N. borealis*. Nests are frequently in holes in mortar and brickwork, although there are a number of records of the species nesting in beetle exit holes in dead wood. The only recorded prey in Britain are psocid nymphs. The adults will visit honeydew (*pers. obs.*).

Occurrence: 4, 4   *v-ix*   1986-2006
National status   Shirt: RDB3   Falk: pRDBK
Kent status   Waite: KRDB1   Present work: pKRDB2

Tribe Trypoxylini. There is only one British genus, *Trypoxylon*, with all five of the indigenous species known from Kent. The *T. figulus* segregates (Pulawski, 1984) are treated on a composite map as well as separately. The species are all aerial nesters and predators of Araneae. Unlike the Pompilidae, several small spiders are provisioned in each cell.

Trypoxylon attenuatum Smith

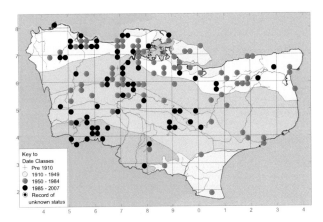

Common and widely distributed across the county, although scarce in coastal areas and fen. The nest is usually excavated in pithy stems and is provisioned with small spiders. A possible host of some elampine Chrysididae.

Occurrence: 61, 140   *v-x*   1900-2007
No status.

Trypoxylon clavicerum Lepeletier & Serville

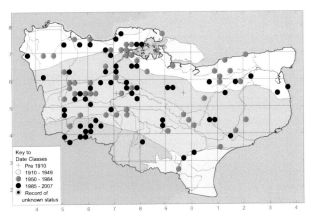

Common and widely distributed across the county, although scarce in coastal areas and fen. The nest is usually made in beetle exit holes in dead wood and is provisioned with small spiders. A possible host of some elampine Chrysididae and *Chrysis gracillima*.

Occurrence: 58, 111   *v-ix*   1897-2007
No status.

Trypoxylon figulus aggregate

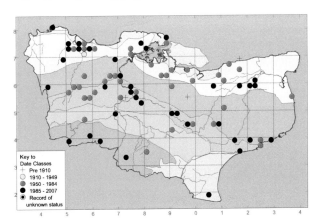

As the species of this aggregate are difficult to identify and the segregates only recently established as good species, a composite map is included here, showing all dots for the three species (*T. figulus* (*s.str.*), *T. medium* and *T. minor*) plus undetermined records.

Trypoxylon figulus (Linnaeus) (*s. str.*)

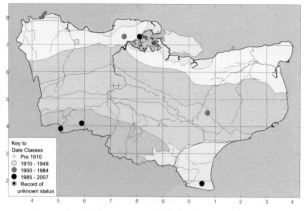

This species is indeed scarce, following the removal of *T. medium* and *T. minor*. However, I hesitate to give it rarity status as it is probably under recorded. It shows a possible decline. The nest is formed in a variety of situations including preformed holes and soft rock cliffs. It is provisioned with small spiders.

Occurrence: 4, 10   *v-viii*   1896-2004
Kent status   Present work: pKa

Trypoxylon medium de Beaumont

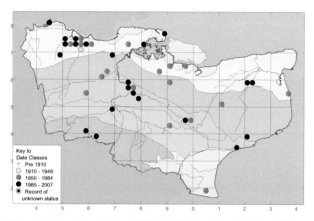

The absence of records pre 1939 may be due in part to the relatively recent separation of this species from *T. figulus* but it is likely to have increased in proportion to that species over the 20th century. The nest is frequently found in beetle exit holes in dead wood and is provisioned with small spiders. A possible host of some elampine Chrysididae.

Occurrence: 22, 41   *v-ix*   1939-2007
No status.

Trypoxylon minor de Beaumont

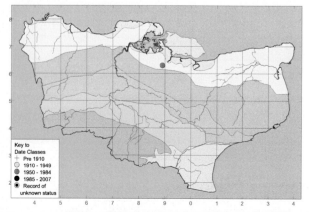

Known from only one British specimen, captured at Sittingbourne, Kent by J C Felton (1988). Whilst there is no reason to question the identification, there have been no further British records and the species remains enigmatic here. The specimen may have been from a nest in a chestnut paling fence post.

Occurrence: 0, 1   *vi*   1959
National status   Falk: pRDBk
Kent status   Waite: KRDBK   Present work: pKRDBK

Tribe Crabronini.

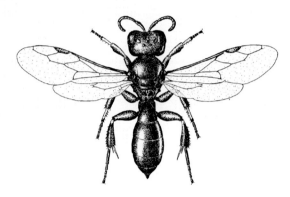

♀ *Crossocerus annulipes*

The British species of Crabronini were once all placed in a blanket genus, *Crabro*. It is, however, a diverse group with six recognised British genera, two of these being large. *Crossocerus* has 20 species recorded from Kent and *Ectemnius*, nine of the ten British species. It is unclear at present if *Crabro scutellatus* is a Kent species.

Many crabronine species are aerial nesters but *Crabro*, some *Crossocerus*, *Lindenius* and *Entomognathus* nest in sandy soil.

The prey are mainly adult Diptera. *Crossocerus* prey additionally includes other insect orders, such as Hemiptera, and *Entomognathus* preys on adult beetles. The prey of *Ectemnius* are predominantly syrphid flies although other dipterous families are taken. The adults of many species are to be found on flowers of Apiaceae.

Crabro cribrarius (Linnaeus)

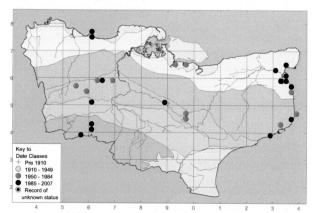

A large, conspicuous species unlikely to be substantially under recorded, hence given scarcity status here. Most records are from sandy soils in which the wasp nests. The prey are Diptera of the families Muscidae (*sensu lato*), Syrphidae and others.

Occurrence: 16, 27   *vi-viii*   1900-2007
Kent status   Present work: pKb

Crabro peltarius (Schreber)

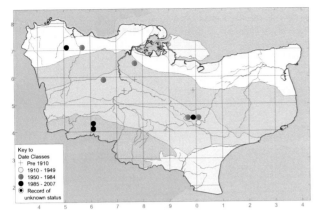

Not as large as the last species but still rather conspicuous, probably not much under recorded. Mostly on sandy soils and a localised species. The nests are excavated in the ground and provisioned with Diptera of the families Muscidae (*sensu lato*), Stratiomyidae and Therevidae.

Occurrence: 4, 13   *v-viii*   1896-2003
Kent status   Present work: pKa

Crossocerus annulipes (Lepeletier & Brulle)

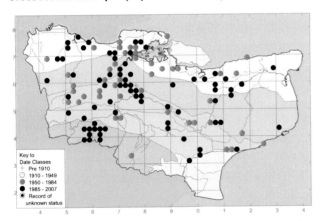

A very common species but not everywhere. Not recorded from marsh and fen areas. This may be due to its habit of nesting in dead wood and possibly pithy stems. Prey are Heteroptera and rarely Psyllidae.

Occurrence: 73, 119   *vi-x*   1890s-2007
No status.

Crossocerus binotatus Lepeletier & Brulle

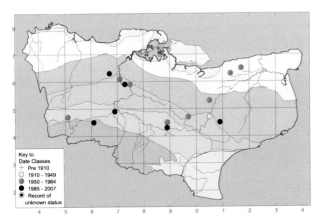

This appears to be a scarce species in Kent but found on most soils types in the county. Nests are excavated in hard dead wood and provisioned with *Rhagio* (Diptera).

Occurrence: 6, 15   *vi-ix*   1946-2006
National status   Falk: pNa
Kent status   Waite: Notable   Present work: pKb

Crossocerus capitosus (Shuckard)

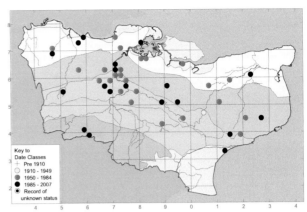

This species is not frequent on the High Weald sands and is rare on the Weald clay. Particularly recorded on the Eocene sands, chalk and Lower Greensand. Nests in the pith of living cut stems and dead wood. Prey are Diptera and occasionally Psyllidae (Hemiptera).

Occurrence: 19, 44   *v-ix*   1946-2007
No status.

Crossocerus cetratus (Shuckard)

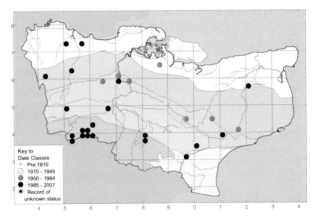

Of rather similar frequency to the last species but most frequent on the High Weald sands. In other areas may be mostly on the scarps. The nest is most often constructed in dead wood but occasionally in stems. Prey are usually Diptera but a few Hemiptera may occasionally be taken.

Occurrence: 21, 28   *v-lx*   1966-2004
No status.

Crossocerus dimidiatus (Fabricius)

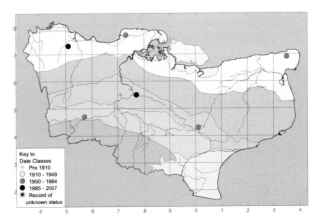

Although there are a number of county records, these are from few squares and the species is given rarity status. There was a long gap without records in the 20th century. Found on chalk and sandy soils. Nests in rotten wood and provisions with Diptera.

Occurrence: 2, 8   *v-viii*   1895-1998
Kent status   Present work: pKRDB2

Crossocerus distinguendus (Morawitz)

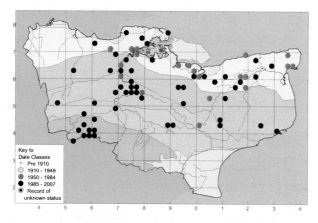

First recorded as British from Kent material. It has, however, rapidly expanded its range in the south-east and no longer requires scarce or threatened status. Nests in sandy soils, in the mortar in walls and beetle exit holes in wooden posts. Prey has been recorded as small Diptera in continental Europe but possibly as aphids in Britain.

Occurrence: 53, 73 *v-ix* 1974-2007
National status Shirt: RDB3 Falk: pNa
No Kent status.

Crossocerus elongatulus (Vander Linden)

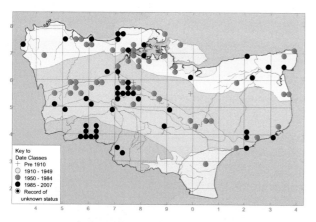

A common species but possibly showing a decline. Frequent on sandy soils but not so on other strata. It also appears to be now scarce in the areas where *C. distinguendus* was first recorded in the county but this may be due in part to observer bias. Nests in similar situations to the latter species and preys on small Diptera, including Drosophilidae.

Occurrence: 46, 96 *iv-ix* 1895-2007
No status.

Crossocerus exiguus (Vander Linden)

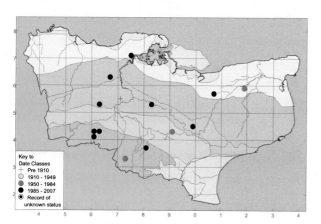

A scarce species formerly considered rare in Kent. Records are well distributed in time and scattered across much of the county. The species is often recorded from coppiced chestnut woodland, in areas showing early re-growth. It is never numerous at a site. The nest is dug in sparsely vegetated, level soil or vertical banks and provisioned with aphids or acalypterate flies.

Occurrence: 10, 14 *vi-ix* 1896-2005
National status Shirt: RDB3 Falk: pRDB3
Kent status Waite: KRDB2 Present work: pKa

Crossocerus megacephalus (Rossius)

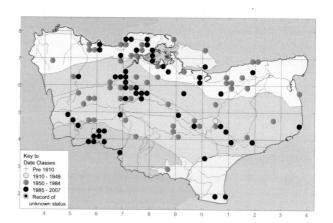

A common and widely distributed species nesting in beetle exit holes in dead wood. It is reported to be sometimes a communal nester, two or more females sharing a common entrance hole. The cells are stocked with various Diptera.

Occurrence: 54, 109 *v-x* 1904-2007
No status.

Crossocerus nigritus Lepeletier & Brulle

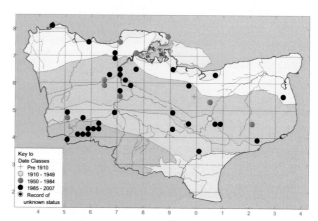

A fairly common species although not as frequent as the last. Recorded across much of the county. Nests in hollow stems and stocks the cells with small Diptera.

Occurrence: 33, 45 *v-viii* 1902-2005
No status.

Crossocerus ovalis Lepeletier & Brulle

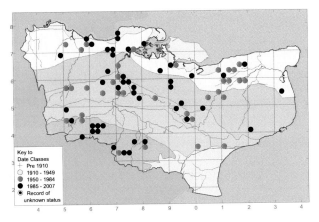

A common species in the county, distributed mainly on sandy soils. It is rarely coastal. Nests in sandy ground and preys on small Diptera.

Occurrence: 45, 80 *v-ix* 1900-2007
No status.

Crossocerus palmipes (Linnaeus)

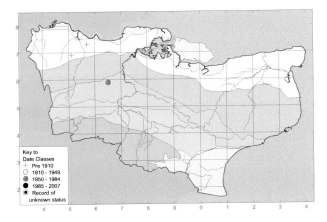

Very rare in the county, although widely distributed in the UK. Most frequent from open, sandy habitats. The female nests in bare or sparsely vegetated sandy soil although a male was taken in a beetle hole in a dead willow (Dr G H L Dicker notebook). The nest is unicellular and provisioned with dipterous flies, e.g. Muscidae, Dolichopodidae, Chloropidae and Lauxaniidae.

Occurrence: 0, 3 *vii-viii* 1859-1982
National status Falk: pNb
Kent status Waite: Notable Present work: pKRDB1

Crossocerus podagricus (Vander Linden)

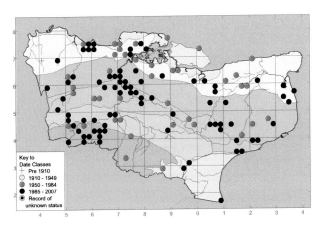

Very common and widely distributed across Kent, on sands, clays and chalk. Nests in hard, dead wood and provisions with small Diptera, particularly Nematocera.

Occurrence: 80, 122 *v-ix* 1890s-2007
No status.

Crossocerus pusillus Lepeletier & Brulle

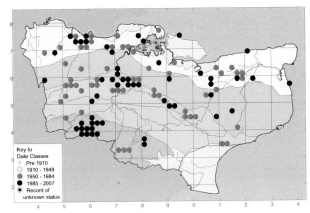

Particularly frequent on sandy soils but scarce on the Weald clay. Occasionally present on chalk scarps. Nests in the soil and preys on small Diptera.

Occurrence: 54, 111 *v-x* 1890s-2007
No status.

Crossocerus quadrimaculatus (Fabricius)

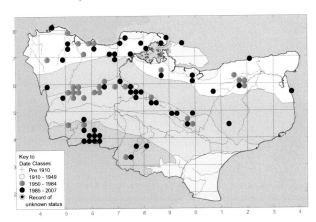

Common on sandy soils, where it frequently nests in the earth root plates of uprooted trees. Prey are mainly Diptera (of many families) but also rarely small Lepidoptera and Trichoptera.

Occurrence: 52, 85 *vi-x* 1894-2005
No status.

Crossocerus styrius (Kohl)

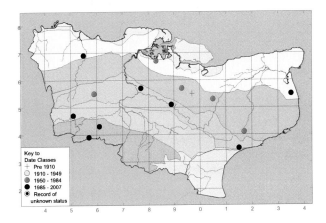

44 The Bees, Wasps and Ants of Kent

A scarce species in Kent but records widely distributed across the county. Apparently, nesting habits and prey are unknown in the UK.

Occurrence: 8, 14 *v-x* 1903-2007
Kent status Present work: pKa

Crossocerus tarsatus (Shuckard)

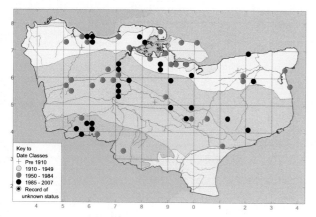

Predominantly on the sand where it is fairly common. Possibly showing a decline. Nests in light soils or mortar in walls and stocks the cells with small Diptera, particularly Empididae.

Occurrence: 27, 62 *v-x* 1896-2007
No status.

Crossocerus vagabundus (Panzer)

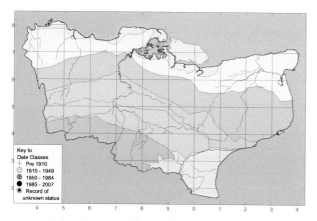

Assumed extinct in the county. The sole known Kent specimen was from Brenchley Gardens, Maidstone (coll. H Elgar). The species nests in dead wood and preys on craneflies (Diptera).

Occurrence: 0, 1 *vi* 1897
National status Shirt: RDB1 Falk: pRDB1+
Kent status Present work: pKRDB1+

Crossocerus walkeri (Shuckard)

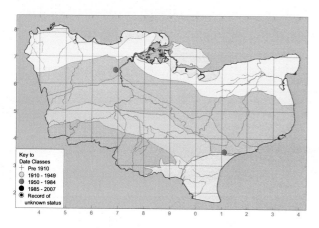

Although widespread in the UK, this species is nowhere common and appears very rare in Kent. There seems to have been a national decline, possibly due to deteriorating water quality and river improvement schemes, which may have affected prey numbers. The female wasp nests in dead wood and provisions the cells with Ephemeroptera, particularly the family Baetidae.

Occurrence: 0, 2 *vii* 1969-1981
National status Falk: pNb
Kent status Waite: Notable Present work: pKRDB1

Crosscoerus wesmaeli (Vander Linden)

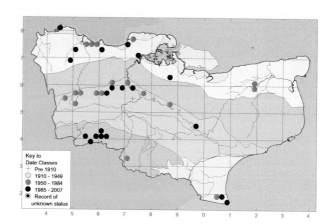

Frequent on sandy soils in the west of the county. Virtually absent from other soil types and scarce in East Kent. Nests in sandy soil and provisions with small Diptera, some nests exclusively with Therevidae.

Occurrence: 20, 40 *v-ix* 1896-2007
No status.

Ectemnius cavifrons (Thomson)
Known in some older literature as "*Crabro cephalotes*".

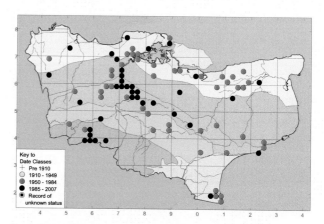

Records scattered across the county, although the species is not frequently recorded on the chalk. The female nests in dead wood and provisions with Diptera, mainly Syrphidae. The adults, as with most *Ectemnius*, are to be commonly found visiting Apiaceae for nectar.

Occurrence: 36, 83 *v-x* 1896-2007
No status.

Ectemnius cephalotes (Olivier)
Synonymy: *E. quadricinctus* (misidentified)

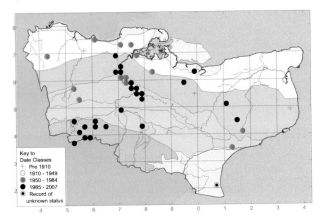

Clearly less common than the last species but not scarce. It may be particularly frequent in the Medway valley. The females may sometimes be communal, two or more sharing a common nest entrance in hard, dead wood. The individual nests within are thought to be separate, however, the females being essentially solitary. The prey is Diptera, particularly Muscidae (*sens. lat.*) and Syrphidae.

Occurrence: 25, 42   vi-x   1850s-2006
No status.

Ectemnius continuus (Fabricius)

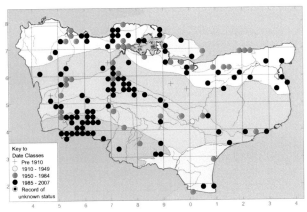

An abundant species most frequently recorded on sandy soils. The female nests in dead wood and preys on Diptera, particularly Muscidae (*sens. lat.*) and Syrphidae.

Occurrence: 110, 144   v-x   1896-2007
No status.

Ectemnius dives (Lepeletier & Brulle)

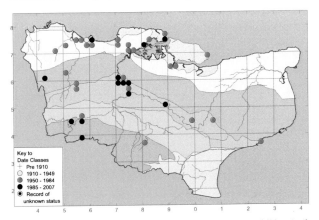

Unknown in the county before 1939 and probably an addition to the fauna since the time of the classical Kent collectors in the late 19th century. The female nests in dead wood and preys on Diptera, particularly Syrphidae and Tachinidae.

Occurrence: 13, 40   v-ix   1939-2007
No status.

Ectemnius lapidarius (Panzer)
Synonymy: *E. chrysostomus*.

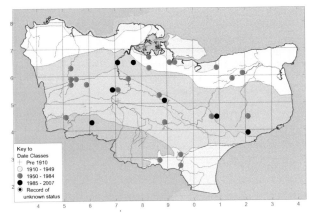

A species that appears to have undergone a serious decline since the 1980s. The female nests in dead wood and preys on Diptera, mainly Syrphidae.

Occurrence: 7, 31   v-ix   1894-2004
Kent status   Present work: pKa

Ectemnius lituratus (Panzer)

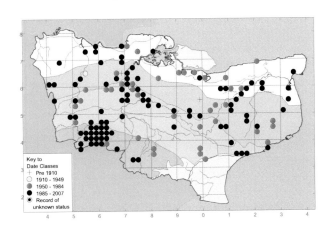

Although very common, this species is infrequently recorded on clay soils. The female nests in dead wood and preys on Diptera, mainly Muscoidea.

Occurrence: 93, 135   v-ix   1899-2007
No status.

Ectemnius rubicola (Dufour & Perris)

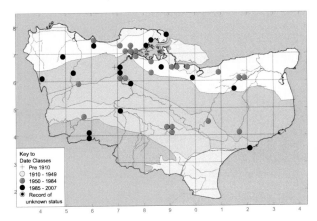

Infrequently recorded since the 1980s and probably never a common species. The female often nests in hollow, pithy stems and occasionally in dead wood, preying on Diptera of various families but often Acroceridae.

Occurrence: 17, 42   v-viii   1902-2006
No status.

Ectemnius ruficornis (Zetterstedt)
Synonymy: E. nigrifrons, E planifrons

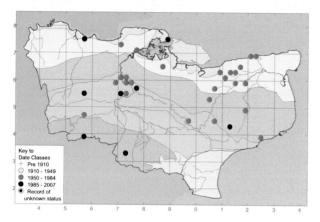

A rather scarce species often associated with open rides in woodland. It is unclear if the species is a fairly recent addition to the county fauna or if its recording is due to increased knowledge. The female nests in dead wood and preys on Syrphidae (Diptera).

Occurrence: 8, 33   v-ix   1969-2001
National status   Falk: pNb
Kent status   Waite: Notable   Present work: pKa

Ectemnius sexcinctus (Fabricius)
Synonymy: E. zonatus, E. saundersi, E. quadricinctus (misidentified)

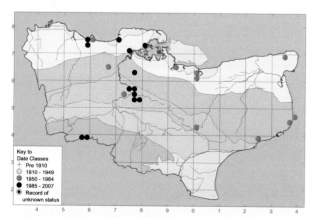

A scarce species sometimes associated with gardens. It is unclear if the species is a fairly recent addition to the county fauna or if its recording is due to increased knowledge. The female nests in dead wood and preys on Diptera: Syrphidae and occasionally Calliphoridae.

Occurrence: 13, 24   vi-ix   1977-2005
National status   Falk: pNb
Kent status   Waite: Notable   Present work: pKb

Lindenius albilabris (Fabricius)

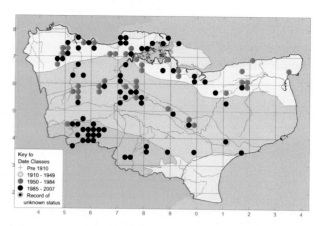

Particularly frequent on sandy soils, scarce on Weald clay. The female nests in sandy soil and provisions with Miridae (Hemiptera) or Chloropidae (Diptera). The adults will visit *Leucanthemum vulgare* for nectar. A host of *Hedychridium coriaceum* (Chrysididae; Elampinae), a species so far not recorded from Kent.

Occurrence: 69, 110   vi-ix   1897-2007
No status.

Lindenius panzeri (Vander Linden)

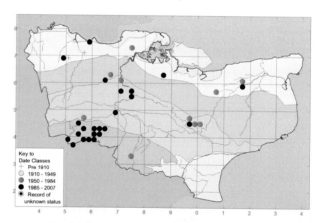

As with the last species, mostly from sandy soils but distinctly scarcer. The female nests in light soils and preys on male Chloropidae (Diptera).

Occurrence: 23, 35   vi-ix   1850s-2007
No status.

Entomognathus brevis (Vander Linden)

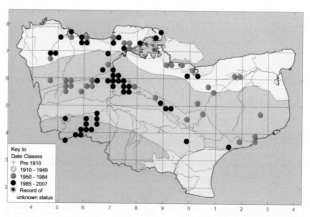

Mainly from sandy soils. The female nests in soil, often sand, and preys on halticine and cryptocephaline Chrysomelidae (Coleoptera).

Occurrence: 47, 85   v-ix   1897-2007
No status.

Rhopalum clavipes (Linnaeus)

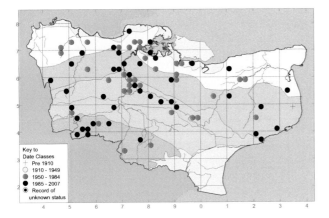

Scarce on clay soils, otherwise quite common but possibly showing a slight decline. Usually nests in hollow, pithy stems or straws but shows some versatility – will occasionally nest in dead wood, old mortar or sandy banks. The prey is mainly barklice (Psocoptera) but occasionally Diptera and Hemiptera.

Occurrence: 41, 77 v-x 1896-2006
No status.

Rhopalum coarctatum (Scopoli)

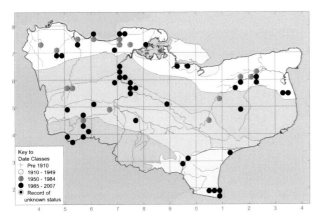

Not confined to any one soil type but often found near water bodies. The female usually nests in stems but occasionally in dead wood. The prey is usually Diptera but occasionally Psocoptera, Hemiptera or Staphylinidae (Coleoptera).

Occurrence: 43, 59 v-ix 1890s-2006
No status.

Tribe Oxybelini.

♀ *Oxybelus uniglumis* carrying green bottle fly prey impaled on the sting   © Jeremy Early

Crabroninae

This is a small, structurally uniform group with one British genus, *Oxybelus*. There are three indigenous species, of which two have been reliably recorded from Kent.

Oxybelus are small species that prey on Diptera and nest in sandy soil. Females of some species carry their prey impaled on the sting.

Oxybelus argentatus Curtis
Synonymy: *O. mucronatus*

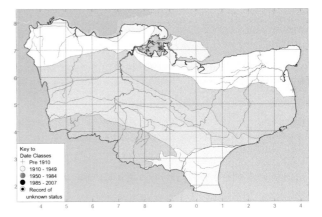

Only known in Kent from one 19th century record, hence is probably extinct in the county. Mainly a species of coastal sand dunes but found on inland heaths in other counties. The female nests in loose, bare sand and preys on *Thereva* (Diptera; Therevidae).

Occurrence: 0, 1 *A Kent flight period cannot be defined* 1877
National status   Falk: pNa
Kent status   Present work: pKRDB1+

Oxybelus uniglumis (Linnaeus)

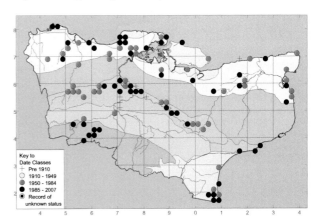

Distribution strongly and positively correlated with sandy soils, both coastal and inland. The female nests in sandy soil and preys on Diptera, mainly Muscoidea.

Occurrence: 50, 101 v-ix 1894-2007
No status.

Subfamily Pemphredoninae

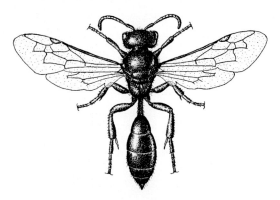

♀ *Mimumesa dahlbomi*

The pemphredonines are predominantly a group of small, black wasps that usually go completely unnoticed by the layperson. The subfamily is split into two tribes, Pemphredonini and Psenini. The former tribe contains a number of species that can prove abundant, whilst the latter clearly does not form a dominant element in the British crabronid wasp fauna. A character found in both tribes is a petiolated waist, although this is not present in *Spilomena* and *Passaloecus*, and is very weakly developed in *Diodontus*.

Tribe Psenini. This tribe is divided into two subtribes, Psenina containing the Kent genera *Mimesa* and *Mimumesa*, and Psenulina, which contains only *Psenulus* in the indigenous fauna. Psenina tend to prey on Cicadellidae (Hemiptera: Auchenorrhyncha) whilst *Psenulus* capture aphids or psyllids. In the Psenina there are four species of *Mimesa* and three of *Mimumesa* found in the county. These two genera were once considered as part of a blanket genus *Psen* and in the past there has been considerable confusion over the identity of the species. Hence, the maps given here are tentative. The nesting habits of some *Mimumesa* need further investigation. There are three *Psenulus* species (Psenulina) in Britain and all are found in Kent.

Mimesa bicolor (Jurine)
At times in the past misidentified as *M. equestris* and confused as *M. rufa*.

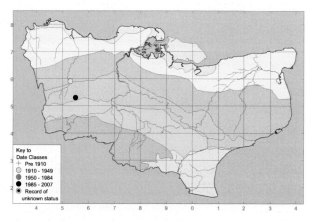

An extremely rare wasp in the county, recently recorded on the Lower Greensand and only known from West Kent. It nests in sandy soil and preys on Cicadellidae (Hemiptera).

Occurrence: 1, 2   *v-vii*   1922, 1998
National status   Shirt: RDB3   Falk: pRDB2
Kent status   Present work: pKRDB1

Mimesa bruxellensis Bondroit

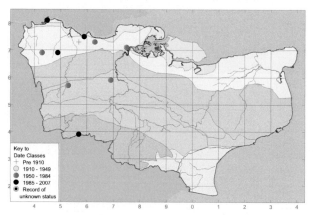

A rare species so far found only in West Kent. Present only on sandy soils where it nests. Prey are Cicadellidae (Hemiptera).

Occurrence: 4, 10   *vi-ix*   1899-2006
National status   Falk: pNa
Kent status   Waite: Notable   Present work: pKRDB3

Mimesa equestris (Fabricius)
At times in the past misidentified as *M. bicolor*.

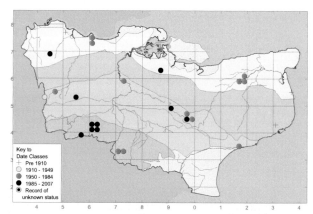

Although widespread on the sands, this is a scarce species in the county. Nests in sandy soils and provisions with Cicadellidae (Hemiptera).

Occurrence: 10, 26   *vi-ix*   1850s-2007
Kent status   Present work: pKb

Mimesa lutaria (Fabricius)
Synonymy: *M. shuckardi*

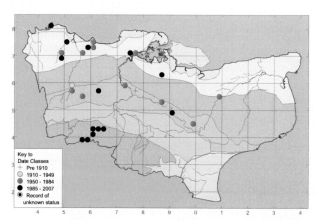

Most frequent in West Kent but records scattered on sandy soils across much of the county. Nests in the sand and mainly preys on Jassinae with occasional Typhlocybinae (Hemiptera: Cicadellidae).

Occurrence: 14, 27   *vi-ix*   1888-2005
No status.

Pemphredoninae

Mimumesa dahlbomi (Wesmael)

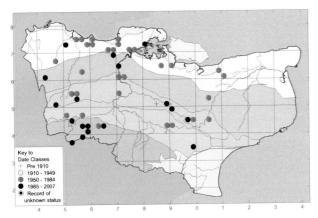

Although fairly common, this species is distributed mainly on sandy soils. It nests in beetle burrows in dead wood and preys on Delphacidae and Cicadellidae (Hemiptera).

Occurrence: 18, 49 *v-ix* 1897-2007
No status.

Mimumesa spooneri (Richards)

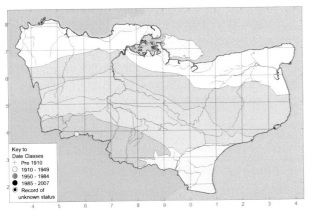

Two old records from the north of the county may have been referable to *M. unicolor* but specimens have not been located. There is a voucher for the one confirmed record. The species may be extinct in the county or present in extremely low numbers. Nest and prey are unknown.

Occurrence: 0, 1 *vii* 1924
National status Shirt: RDB3 Falk: pRDB3
Kent status Present work: pKRDB1

Mimumesa unicolor (Vander Linden)

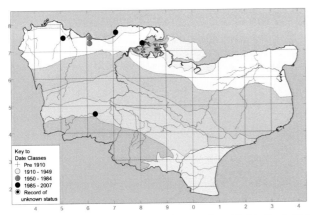

A rare species in the county and only recently added to the British list (Else & Felton, 1994). Nationally, many records are from coastal soft rock cliffs or estuaries. There appears to be an association with *Phragmites* beds but it is likely that the species nests in the soil. The prey is unknown.

Occurrence: 4, 6 *vi-ix* 1983-2005
National status Shirt: RDB3 Falk: pNa
Kent status Waite: Notable Present work: pKRDB3

Psenulus concolor (Dahlbom)

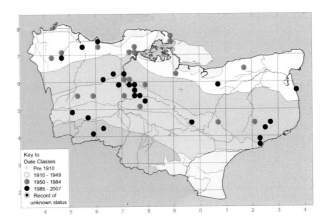

Records scattered across much of the county but the species is rather scarce on clay soils. Nests in pithy stems and preys on Psyllidae (Hemiptera).

Occurrence: 24, 44 *v-viii* 1896-2007
No status.

Psenulus pallipes (Panzer)
Synonymy: *P. atratus*

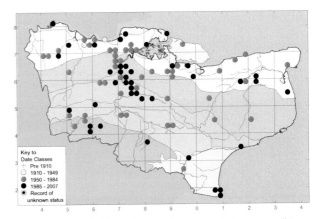

A common wasp in Kent. Recorded from on all the main soil types including the weald clay but with a tendency to be scarce on the chalk. The female nests in pithy stems or beetle borings in dead wood, stocking the cells with aphids (Hemiptera).

Occurrence: 40, 90 *v-ix* 1914-2006
No status.

Psenulus schencki (Tournier)

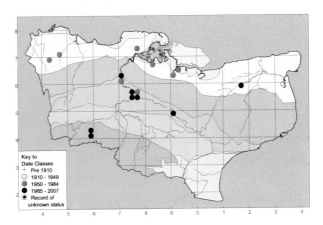

50 The Bees, Wasps and Ants of Kent

Added to the British list in the middle of the 20th century (Yarrow 1954) and only known in Britain from the south-east. A very scarce species in the county, apparently most frequent on sandy soils. Nests in pithy stems or decayed wood, the cells being provisioned with Psyllidae (Hemiptera).

Occurrence: 8, 16   *v-ix*   1970-2007
National status   Shirt: RDB3   Falk: pNa
Kent status   Present work: pKa

Tribe Pemphredonini. This group is formed of two sub-tribes, Pemphredonina and Stigmina, that differ in the area of the pterostigma of the fore wing relative to that of the marginal cell. Aphids (Hemiptera) are the principal prey of the tribe but *Spilomena* usually capture thrips (Thysanoptera). There are two British and Kent genera of Stigmina, *Stigmus* and *Spilomena*, with two and four species respectively. The species are all aerial nesters. *Spilomena* comprises minute wasps, little known and easily overlooked. Indeed, most species in this subtribe may be under recorded and I have no firmly identified data for *Spilomena* before 1964. The British genera of Pemphredonina are *Pemphredon*, *Diodontus* and *Passaloecus*. 17 of the 18 British species are recorded from Kent, apart from *Passaloecus monilicornis*. The taxonomy of the British *Passaloecus* was put on a firm footing by Yarrow (1970), with additions by Richards (1980) and Guichard (2002).

Spilomena beata Bluethgen

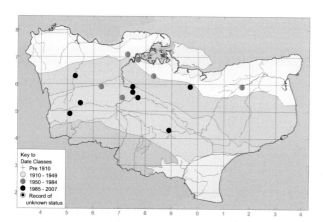

A rather scarce species in the county; apparently not recorded from the High Weald sands. Some records may be from scarp faces. Nests in holes in wood and possibly pithy stems. The prey has not definitely been recorded in Britain.

Occurrence: 8, 14   *vi-viii*   1964-1998
Kent status   Present work: pKa

Spilomena curruca (Dahlbom)
Synonymy: *S. differens*

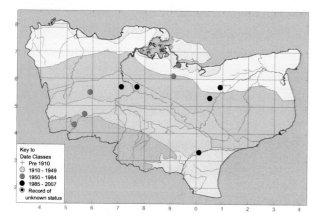

Rare according to the known data but probably very under recorded. Mainly from on sandy strata. Nests in beetle holes in wood and preys on thrips (Thysanoptera). The adults have been found visiting *Mercurialis* sp. in numbers.

Occurrence: 5, 10   *vii-ix*   1964-1997
Kent status   Present work: pKa

Spilomena enslini Bluethgen

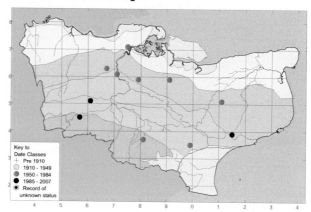

There are not enough distributional data to assess any soil type preferences and the species may be very under recorded. Nests in hollow pithy stems and preys on thrips.

Occurrence: 3, 11   *vi-ix*   1979-1991
Kent status   Present work: pKa

Spilomena troglodytes (Vander Linden)
Synonymy: *S. vagans*

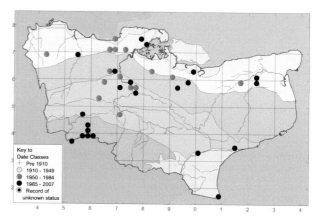

Scarcer on clay and chalk soils but the species is presently classified as common because it is undoubtedly frequently over looked. The female nests in hollow stems and beetle borings in wood. Preys on thrips. Adults have been found visiting *Solidago* sp. (*pers. obs.*).

Occurrence: 22, 37   *vi-ix*   1975-2007
No status.

Stigmus pendulus Panzer

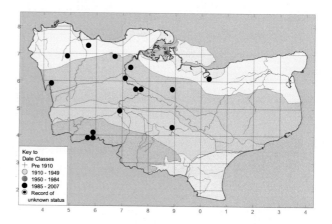

First recorded as British from Kent. A local species apparently not restricted to any particular soil type. Nests in dead wood and pithy stems, provisioning with aphids.

Occurrence: 15, 15 *v-ix* 1986-2007
National status Falk: pRDBK
Kent status Waite: KRDB2 Present work: pKb

Stigmus solskyi Morawitz

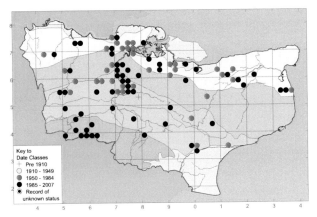

A common species, most frequently recorded on sandy soils. Nests in dead wood and pithy stems, provisioning with aphids.

Occurrence: 55, 92 *v-ix* 1893-2007
No status.

Pemphredon austriaca (Kohl)
Synonymy: *P. enslini*

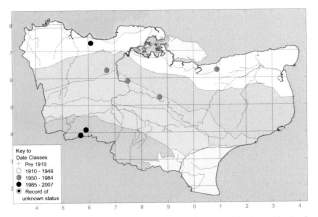

A rare species possibly absent from the weald clay. Nests in brambles and probably other pithy stems. The prey are aphids.

Occurrence: 3, 7 *vi-viii* 1980-2002
National status Shirt: RDB3 Falk: pRDB3
Kent status Waite: KRDB2 Present work: pKRDB2

Pemphredon inornata Say
Synonymy: *P. shuckardi*

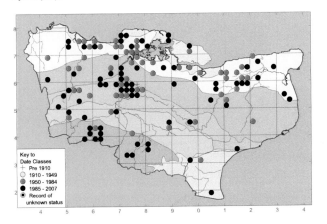

A very common species but less frequently recorded on chalk and clay soils. Nests in pithy stems and preys on aphids.

Occurrence: 73, 130 *v-x* 1893-2007
No status.

Pemphredon lethifer (Shuckard)

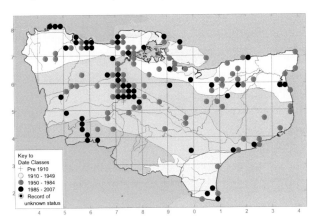

Most frequently recorded on sandy soils. The occurrence figures are inflated by the rearing records of Dr G H L Dicker but nevertheless a very common species. Nests in pithy stems and preys on aphids.

Occurrence: 52, 137 *v-ix* 1897-2007
No status.

Pemphredon lugubris (Fabricius)

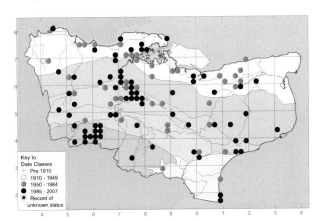

The largest British species of the genus, hence probably more completely recorded than *P. inornata* and *P. lethifer*. Nests in dead wood, perhaps most often in beetle exit holes, and preys on aphids.

Occurrence: 70, 114 *v-xi* 1894-2007
No status.

Pemphredon morio Vander Linden
Including *P. clypealis*. Synonymy: *Ceratophorus morio*.

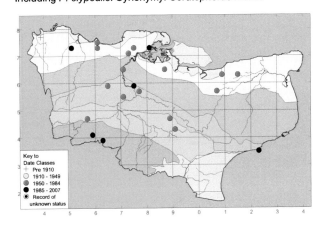

52 The Bees, Wasps and Ants of Kent

A scarce species, possibly most frequently recorded from scarp faces in the county. Nests in rotten wood, prey aphids.

Occurrence: 6, 22 *v-ix* 1897-2007
National status Shirt: RDB3 Falk: pNa
Kent status Waite: Notable Present work: pKa

Pemphredon rugifer (Dahlbom)
Synonymy: *P. mortifer, P. wesmaeli.*

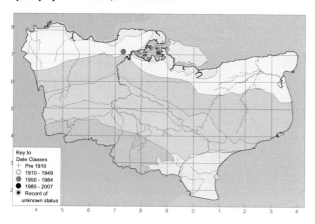

A very rare species in the county for which no biotope preferences can be given. Nests in dead wood and presumably preys on aphids.

Occurrence: 0, 2 *vii-viii* 1924, 1975
National status Shirt: RDB3 Falk: pRDB3
Kent status Waite: KRDB1 Present work: pKRDB1

Diodontus insidiosus Spooner

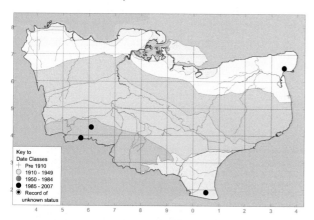

A very rare species recorded in the county from sands and coastal shingle. Nests in sandy soils and preys on aphids.

Occurrence: 4, 5 *vi-viii* 1891-2006
National status Shirt: RDB3 Falk: pRDB3
Kent status Present work: pKRDB2

Diodontus luperus Shuckard

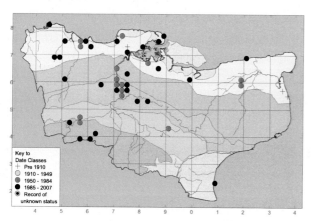

Mostly from the sands in West Kent. Nests in sandy soils and preys on aphids.

Occurrence: 27, 45 *v-x* 1890s-2006
No status.

Diodontus minutus (Fabricius)

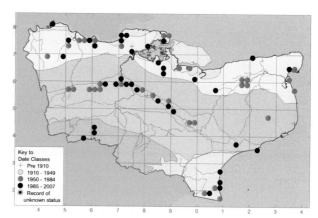

Although common, the dot map suggests that this species exhibits strong preferences on which types of sandy strata it occurs. Nests in sand and preys on aphids.

Occurrence: 36, 72 *v-x* 1897-2007
No status.

Diodontus tristis (Vander Linden)

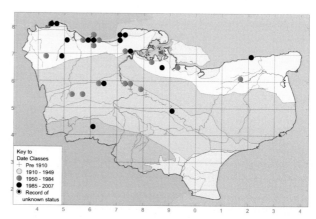

Rare on the High Weald sands and absent from the Dungeness shingle but it is not clear if these are artifacts of recording. Most frequent on the Eocene sands in the north of the county. Nests in sandy soil and occasionally in the mortar of walls. Prey are aphids.

Occurrence: 15, 32 *v-x* 1893-2007
No status.

Passaloecus clypealis Faester

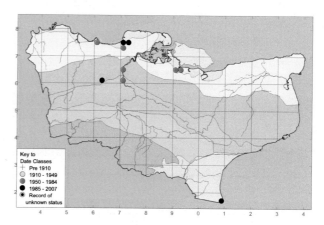

Strongly associated with *Phragmites* beds, where it nests in the hollow reeds themselves or, more often, in the galls of the dipterous fly, *Lipara lucens*. The adults are seldom observed in the field and most records are from rearing or general sweep netting of the *Phragmites*. The prey are assumed to be aphids.

Occurrence: 4, 10 *vi-viii* 1978-1998
National status Shirt: RDB2 Falk: pRDB3
Kent status Waite: KRDB3 Present work: pKRDB3

Passaloecus corniger Shuckard

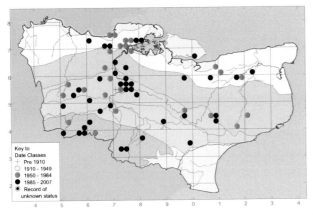

Present on all of the main strata, although infrequently recorded from on the chalk. Rarely coastal. Nests in dead wood and *Lipara* galls. Females apparently steal their aphid prey from the nests of *Psenulus pallipes* and other *Passaloecus* spp. and have not been observed hunting elsewhere.

Occurrence: 40, 66 *vi-ix* 1896-2006
No status.

Passaloecus eremita Kohl

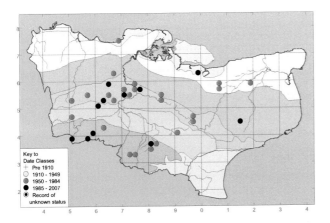

Added to the British list by Richards (1980). Dr G H L Dicker recorded the species frequently, having recognised the jizz of the nest in the field. The nests are often in the bark of standing *Pinus sylvestris* or in dead wood nearby and are uniquely sealed with whitish pine resin. Whilst active there is often a ring of drops of pine resin around the entrance, but of unknown function. The prey are aphids of the family Lachnidae, which are found on pine.

Occurrence: 11, 35 *v-x* 1979-2004
National status Shirt: RDB3 Falk: "No longer believed native, therefore no status."
Kent status Waite: KRDBK Present work: pKb

Passaloecus gracilis (Curtis)

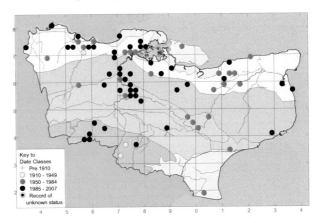

Scarce on the chalk and weald clay but recorded quite frequently on the London Clay in the Blean; otherwise common on the sands. Nests in beetle burrows in dead wood, in pithy stems or in the burrows of *Rhyacionia* (Lepidoptera; Tortricidae). Prey are aphids, usually hunted on low herbage.

Occurrence: 44, 73 *v-ix* 1897-2007
No status.

Passaloecus insignis (Vander Linden)

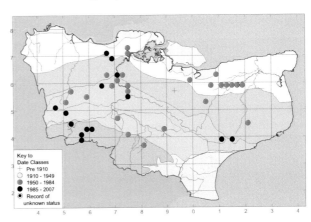

Modern data possibly with a southern and western bias in the county. Nests in pithy stems or beetle borings in dead wood. Preys on aphids.

Occurrence: 14, 40 *vi-ix* 1890s-2007
No status.

Passaloecus singularis Dahlbom

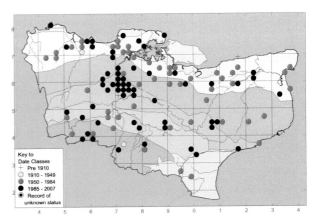

A very common species most frequently recorded on sandy soils. Frequently reared by Dr G H L Dicker from bramble stems, possibly biasing the occurrence data. Nests have only been recorded from pithy stems. Cell partitions are uniquely made of resin, mud and small stones. Prey are aphids.

54 The Bees, Wasps and Ants of Kent

Occurrence: 58, 114 *v-ix* 1896-2007
No status.

Passaloecus turionum Dahlbom

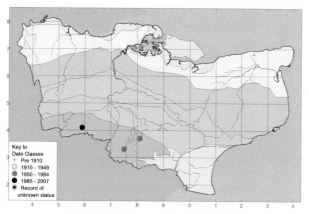

Only recently distinguished in the British fauna (Guichard, 2002), previously confused under the name *P. gracilis* with that species (p. 53). So far only recorded on the weald sands in Kent. Reared by others from *Pinus sylvestris*, and a specimen was also collected entering a burrow in that tree species. A female was captured in the county on a gate post, possibly suggesting that the species also nests in other dead wood. Prey are arboreal aphids.

Occurrence: 1, 3 *vi,viii* 1984, 2002
Kent status Present work: pKRDB1

Subfamily Bembicinae (= Nyssoninae)

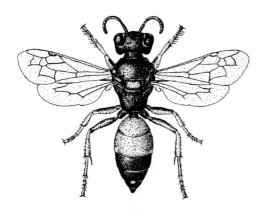

♀ *Harpactus tumidus*

This is a large, diverse subfamily world wide but is poorly represented in Britain. The tribes Stizini and Heliocausini are not found here and Bembicini is known only from the Channel Islands: *Bembix rostrata*. The British and Kent forms are in the genera *Didineis* (Alyssonini), *Gorytes, Lestiphorus, Harpactus, Argogorytes* (Gorytini) and *Nysson* (Nyssonini). The tribe Gorytini appears to be polyphyletic if the cladograms of Melo (1999) are correct. The "cactoid" diagram of Bohart and Menke (1976) implies the same.

The species are all "digger wasps", i.e. they excavate nest burrows in the ground, or are cleptoparasites evolutionarily derived from them. Most of the predacious species nest in sandy soils.

Tribe Alyssonini. There is only one British species, *Didineis lunicornis*, which is found in Kent.

Didineis lunicornis (Fabricius)
Synonymy: *Alysson lunicornis*

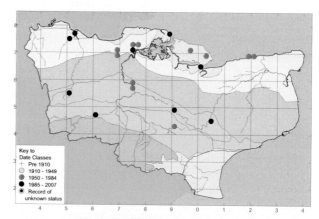

A scarce species which can be locally frequent - often to be found in damp, shaded places on clay soils. The females prey on Cicadellidae and Delphacidae (Auchenorrhyncha) which are brought to nests typically dug close to water bodies. The females run a little like pompilids whilst hunting. Males have been recorded on Apiaceae. Parasites have not been recorded in Britain.

Occurrence: 9, 20 *vii-ix* 1972-2005
National status Shirt: RDB3 Falk: pNa
Kent status Present work: pKa

Tribe Nyssonini. Three of the four British species of *Nysson* have been recorded from Kent, the exception being *N. interruptus*. The species are cleptoparasites of Gorytini.

Nysson dimidiatus Jurine

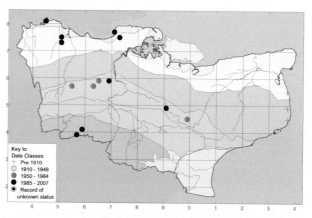

A scarce species in the county, confined to sandy soils. A cleptoparasite of *Harpactus tumidus*, a more widespread species.

Occurrence: 9, 16 *vi-ix* 1850s-2007
National status Falk: pNb
Kent status Waite: Notable Present work: pKa

Nysson spinosus (Foerster)

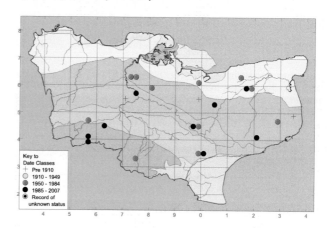

This is a scarce species in Kent and possibly undergoing a decline. Found on chalk, sand and clay soils. A known cleptoparasite of *Argogorytes* species and *Gorytes quadrifasciatus*.

Occurrence: 9, 23 v-vii 1898-1999
No status.

Nysson trimaculatus (Rossius)

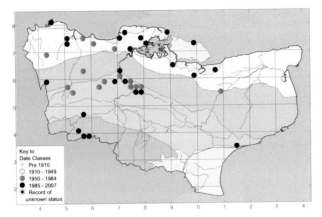

Although here classified as widespread, the distribution of the species is mainly on sedimentary sands. Occasionally found on alluvial sands and there are few dots on the chalk and London clay. There is a strong westerly bias to the data. A cleptoparasite of *Lestiphorus bicinctus* particularly and occasionally, *Gorytes quadrifasciatus*. The species is recorded more regularly than either host, possibly due to its habit of sunning on broad leaves.

Occurrence: 24, 39 vi-ix 1964-2007
National status Falk: pNb
Kent status Waite: Notable   Present work: No Kent status.

Tribe Gorytini. The four genera and all of the British species are found in Kent, but none are particularly common. The species are mainly predators of frog hoppers (Auchenorrhyncha: Cercopidae). *Argogorytes* are said to prey on nymphs of the frog hoppers, extracting one from its "spittle" and stinging it, then carrying to the nest.

Lestiphorus bicinctus (Rossius)
Synonymy: *Gorytes bicinctus*

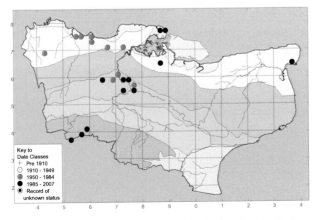

A rather local species in the county found almost exclusively on sandy soils, some possibly alluvial. A predator of cercopids (Hemiptera), particularly *Philaenus* and *Neophilaenus* adults. The main host of *Nysson trimaculatus* in Kent.

Occurrence: 12, 22 vii-ix 1975-2005
National status Falk: pNb
Kent status Waite: Notable   Present work: pKb

Pemphredoninae to Bembicinae 55

Gorytes laticinctus (Lepeletier)

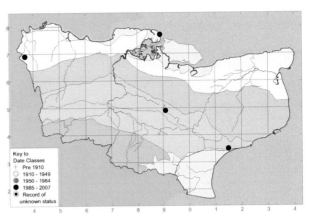

A very rare species in the county and also rare nationally. Known from sandy soils, sedimentary and alluvial. A predator of adult Cercopidae, including particularly *Philaenus* but also *Cercopis* and *Aphrophora* (Aphrophoridae).

Occurrence: 4, 4 v-ix 1989-2005
National status Shirt: RDB3 Falk: pRDB3
Kent status Waite: KRDB1   Present work: pKRDB2

Gorytes quadrifasciatus (Fabricius)

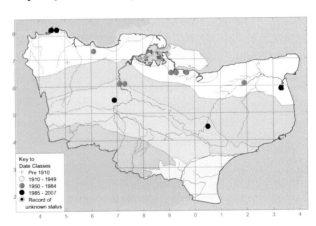

Although not given any national status, this is clearly a scarce species in Kent. The species is distributed mainly on sandy strata in the county, although not known from the Kent high weald. The species is a predator of adult Cercopidae, particularly *Philaenus spumarius*. Host of *Nysson trimaculatus*, *N. spinosus* and, in other southern counties, *N. interruptus*.

Occurrence: 5, 13 vi-ix 1830s-2001
Kent status Present work: pKa

Harpactus tumidus (Panzer)
Synonymy: *Gorytes tumidus, Dienoplus tumidus*.

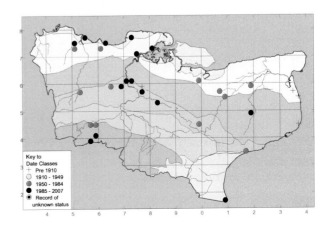

I have not given this species scarcity status in the belief that it is very under recorded. It was found regularly between 1977 and 1985 by Dr G H L Dicker, who clearly recognised the jizz of the species in the field. The distribution is on the sands, including coastal and alluvial. It is a predator of Cercopidae, including *Cercopis*, and possibly *Philaenus* nymphs as well as adults. The usual host of *Nysson dimidiatus* and a possible host for *Hedychridium roseum*.

Occurrence: 15, 31 *vi-ix* 1890s-2007
No status.

Argogorytes fargei (Shuckard)
Synonymy: *Gorytes campestris* (misidentified)

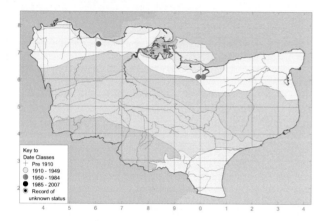

A very rare species in the county, although possibly under recorded and only known on Eocene sand beds. The small number of records probably mean that the full flight period is not completely represented. The prey are nymphs of the frog hopper *Philaenus* (Cercopidae) which are apparently extracted from the "spittle" before being stung and carried to the nest. Host of the cleptoparasite *Nysson interruptus*, which is not recorded from Kent, and possibly *N. spinosus*.

Occurrence: 0, 3 *vi-vii* 1979-1980
National status   Shirt: RDB3   Falk: pNa
Kent status   Waite: Notable   Present work: pKRDB1

Argogorytes mystaceus (Linnaeus)
Synonymy: *Gorytes campestris*

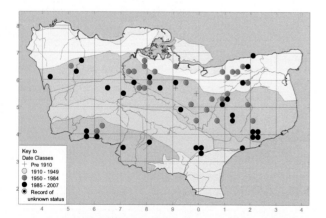

The most frequently recorded Kent gorytine. A species of broad leaved woodland rides and edges, and fairly common in this habitat. Found on most strata but scarce on the weald clay. Like the previous species, a predator of *Philaenus* nymphs and the predatory behaviour is similar. The principle host of *Nysson spinosus* and possibly also parasitised by *N. interruptus*.

Occurrence: 30, 61 *v-vii* 1896-2007
No status.

Subfamily Philanthinae

This subfamily is divisible into two British tribes, Cercerini and Philanthini, each with one genus in Kent and Britain. The species all dig nest burrows in the ground, usually in sandy soil.

Tribe Cercerini.

♀ *Cerceris rybyensis* carrying *Halictus rubicundus* prey
© Jeremy Early

There is only one British genus, *Cerceris*, with six recorded species. One of these, *C. sabulosa*, is probably not native, having been recorded only once in Britain: a female collected near the Kent coast in the mid-nineteenth century. The remaining five species are found with varying frequency in Kent, one probably being extinct. The prey are adult insects, either beetles or bees, depending on species. *Cerceris* can be frequent on some Asteraceae and the umbels of Apiaceae.

Cerceris arenaria (Linnaeus)

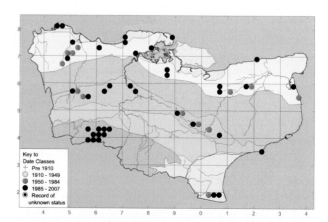

Almost entirely on sandy soils, as implied by the species name. Can be locally abundant in the right conditions. The species is a predator of adult weevils (Curculionidae), including *Otiorrhynchus*. Occasionally, other Coleoptera are used. Several genera of Miltogramminae (Diptera, Sarcophagidae) are cleptoparasites and the chrysidid wasp *Hedychrum niemelai* is a parasitoid.

Occurrence: 39, 54 *vi-ix* 1890s-2007
No status.

Cerceris quadricincta (Panzer)

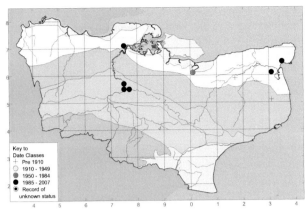

It appears that Kent is a stronghold in Britain for this species but even here it is rare. The Upnor population is under the threat of development. Most records are from the sands but there are some from coastal soft rock cliffs. Work for this atlas found males flying in mid-June, very early for the species. The prey are weevils such as *Sitona lineatus* and *Ceutorhynchus* species (M Edwards, *pers. comm.*). A probable host of *Hedychrum niemelai*.

Occurrence: 6, 10   *vi-ix*   1850s-2007
National status   Shirt: RDB1   Falk: pRDB1
Kent status   Waite: KRDB2   Present work: pKRDB2

Cerceris quinquefasciata (Rossius)
Synonymy: *C. interrupta*

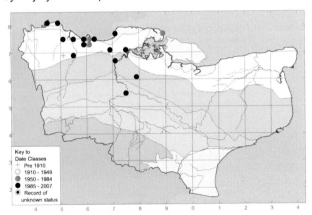

A very scarce species in the county: predominantly from the east Thames corridor although occasionally found in the Medway valley. The prey are weevils (Curculionidae and Apionidae) and sometimes *Meligethes* (Nitidulidae). The species is a probable host of *Hedychrum niemelai*.

Occurrence: 14, 18   *vi-ix*   1850s-2007
National status   Shirt: RDB3   Falk: pRDB3
Kent status   Waite: KRDB3   Present work: pKa

Cerceris ruficornis (Fabricius)
Synonymy: *C. labiata, C. cunicularia*

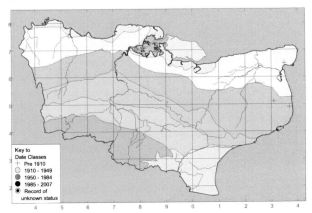

Bembicinae to Philanthinae   57

I have now revised my opinion on this wasp: it did once occur in the county but is now almost certainly extinct (specimens in the Maidstone Museum collection). It is typical of sandy heath whilst the Kent data are mostly from the chalk. The prey are small weevils and no parasites have been recorded in the county.

Occurrence: 0, 3   *viii*   1850s-1900
Kent status   Present work: pKRDB1+

Cerceris rybyensis (Linnaeus)
Synomym: *C. ornata*

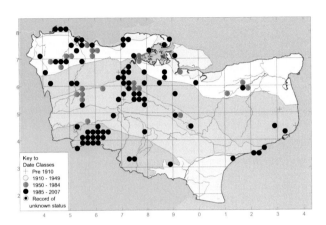

An abundant species, less restricted to the sands than *C. arenaria*: known from chalk heath and rarely on clay soils. It may be that soil texture is a significant factor in choice of nest site. The prey are small bees of several genera but particularly *Halictus*, *Lasioglossum* and small *Andrena*. *Hedychrum niemelai* is a probable parasitoid and miltogrammine flies (Diptera, Sarcophagidae) may be cleptoparasites.

Occurrence: 81, 109   *v-ix*   1896-2007
No status.

Cerceris sabulosa (Panzer)
Synonymy: *C. emarginata*

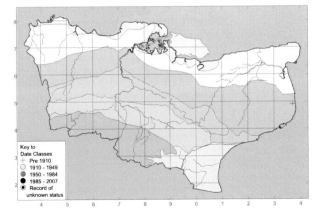

This species is represented from Britain by only one specimen, a female taken at Kingsdown in the nineteenth century. This may represent a vagrant from the continent, as there has never been another British capture. The female preys on small bees as the above species, but possibly incorporates more *Hylaeus* in its hunting. No parasites were recorded from Britain.

Occurrence: 0, 1   *viii*   1861
National status   Shirt: Appendix   Falk: Appendix
Kent status   Present work: Probably not native, therefore no status.

Tribe Philanthini. There is only one British and Kent species, *Philanthus triangulum*.

Philanthus triangulum (Fabricius)

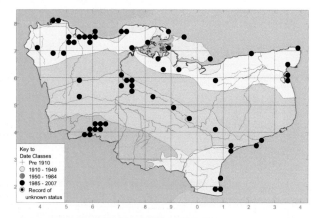

This is a species which undergoes cyclical periods of abundance but has never been so frequently recorded in Britain as in the last 10 years. This is reflected in the no status comment below. May be locally abundant on the sand. A predator principally of the honey bee, *Apis mellifera*, but will rarely take other bees. Host of the extremely rare, non-Kentish, *Hedychrum rutilans*.

Occurrence: 58, 58   vii-ix   1850s-2007
National status   Shirt: RDB2   Falk: pRDB2
No Kent status.

Family Apidae

Worker *Apis mellifera*

This family comprises the bees, mostly hairy aculeates that provision their nests with pollen moistened with nectar. (In the honeybee, glandular secretions of the workers, "brood food", in part replaces the pollen in the larval diet). A few genera are cleptoparasitic; i.e. the female parasite lays her eggs in the nests of other bees. In some of these genera, the adult destroys the host egg whilst in most, the young parasitic larva is equipped with awesome mandibles for this task. In these cleptoparasites, the pollen-gathering and carrying hairs in the adult are usually reduced, sometimes considerably. There are parasites in the bumbles in which the female takes over the nest of other *Bombus* species, destroying the host queen or deposing her as the alpha or dominant individual. These social parasites are workerless. They were formerly placed in a genus *Psithyrus* but are now included in *Bombus*, often as a subgenus equal in rank to the other subgenera.

In some bee species, the female specialises in collecting pollen from one or a very few plant species, sometimes to the point where structural modifications for the specialisation, in both bee and plant, have occurred. Most usually, preference is for a few plant families in an individual bee species. In the social forms however, the life cycle includes a long flight period and such species usually forage on the flowers of a wide variety of plants.

The bees are broadly divisible into two groups, the short-tongued and long-tongued species. The short-tongued bees in the UK comprise the subfamilies Colletinae, Andreninae, Halictinae and Melittinae. There are two long-tongued apid subfamilies, Megachilinae and Apinae.

The plant visit data presented here for Kent owe much to the meticulous record keeping of Dr Gerald Dicker in his unpublished notebooks.

Subfamily Colletinae

The colletines form what was once regarded as the most primitive bee group, based on the structure of the tongue. It appears that Melittinae may in fact be the basal group (S P M Roberts, *pers. comm.*) The small size and comparative hairlessness of *Hylaeus* may be a reversal in evolution.

The British species fall into two tribes, Colletini and Hylaeini, each with a single indigenous genus, *Colletes* and *Hylaeus* respectively. The species tend to fly later in the season, i.e. summer and/or early autumn.

Tribe Colletini.

♀ *Colletes fodiens*

The genus *Colletes* differs from other (non-British) Colletini in having no preapical or apical fimbria on the gaster and differs from all other bees in the sigmoidally outward curving second recurrent vein of the forewing.

Most *Colletes* forage on only a few plant families, including Asteraceae, Ericaceae and Araliaceae; each species usually confines itself to one of these families. They are mining bees, some species nesting in level, sandy ground, others in banks. *C. daviesanus* will additionally burrow into the soft mortar of old walls, perhaps a substitute for sandstone cliff faces. *C. hederae* is a recent addition to the Kent fauna.

Colletes is the host of the cleptoparasitic bee genus *Epeolus* (Apinae).

Colletes daviesanus Smith

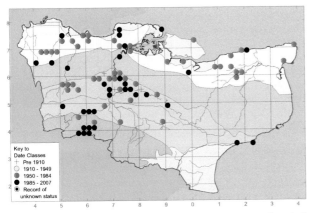

A common bee most frequent on sandy soils, scarce on the chalk. The female nests in flat, sandy soil or banks, or in the soft mortar of old walls. The pollen used to stock the cells comes from Asteraceae such as mayweeds, *Tanacetum* spp., *Senecio*, *Achillea* and *Solidago*. A host of the cleptoparasitic bee, *Epeolus variegatus*.

Occurrence: 33, 84 v-ix 1892-2007
No status.

Colletes fodiens (Geoffroy in Fourcroy)

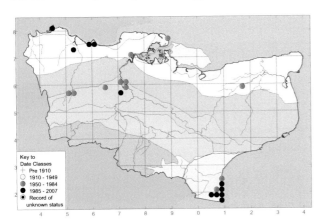

A scarce bee in the county; mostly on sandy soils and frequently coastal. The female forages on Asteraceae: *Tanacetum*, *Achillea* and *Senecio*. A host of *Epeolus variegatus*.

Occurrence: 11, 22 vi-viii 1979-2004
Kent status   Present work: pKa

Colletes halophilus Verhoeff

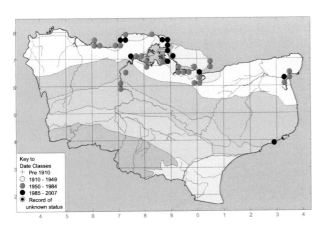

This was thought to be a Doggerland species, now also known from the French coast south to the Spanish border (S P M Roberts, *pers. comm.*). It appears that south-east England holds a significant and important proportion of the total world population. The species is well recorded in Kent, being virtually confined to saltmarsh. The nests are dug in small aggregations in the saline soils of the foreshore and the species may be particularly susceptible to sea level changes. The female predominantly forages on sea aster, *Aster tripolium*, but when that plant is not in flower, will visit *Picris* and other Asteraceae. The host of a large form of the cleptoparasitic bee, *Epeolus variegatus*.

Occurrence: 12, 40 vii-x 1960-2007
National status   Falk: pNa
Kent status   Waite: Notable   Present work: pKa

Colletes hederae Schmidt & Westrich

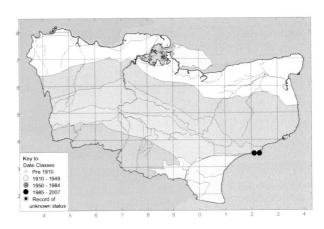

Recently discovered in the county and present in the UK only since 2001. When established, can nest in aggregations numbering thousands, usually in sandy soils. The main forage plant is ivy, *Hedera helix*, hence the late flight period, but it will also forage less commonly on Asteraceae. So far no *Epeolus* have been found as parasites in the UK but *E. cruciger* is known on the continent.

Occurrence: 2, 2 x 2006
Kent status   Present work: pKRDBK

Colletes marginatus Smith

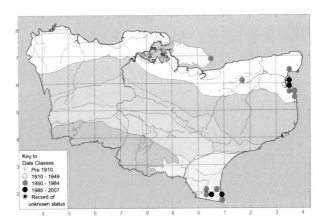

A rare species in the county. The modern distribution is entirely on coastal sands and shingle. The ecology is not well known but the females apparently nest in small aggregations, preferring loose sand. Forage plants are unclear but include *Rubus fruticosus* agg. and probably other species. Host to a small form of *Epeolus cruciger*.

Occurrence: 4, 16 vi-viii 1850s-2003
National status   Shirt: RDB3   Falk: pNa
Kent status   Waite: Notable   Present work: pKRDB3

Colletes similis Schenck

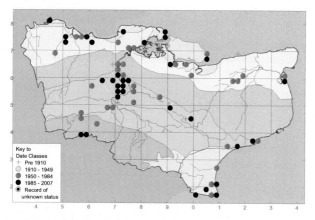

A common bee most frequent on sandy soils including coastal sands; scarce on chalk and clay. The female forages on Asteraceae such as *Leucanthemum vulgare*, *Senecio* spp. and mayweeds. A host of *Epeolus variegatus*.

Occurrence: 32, 69   vi-ix   1897-2007
No status.

Colletes succinctus (Linnaeus)

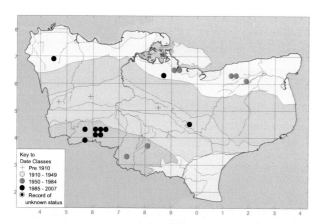

Although common on sandy heaths in other counties this is clearly a scarce species in Kent, where heathy habitats are infrequent. The female forages almost entirely on *Calluna* and *Erica*. A host of the cleptoparasitic bee, *Epeolus cruciger*.

Occurrence: 10, 20   vii-ix   1895-2007
Kent status   Present work: pKa

Tribe Hylaeini.

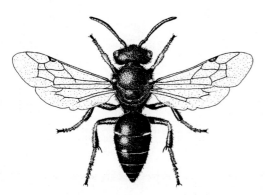

♂ *Hylaeus hyalinatus*

Hylaeus, the only British genus, comprises generally small black, rather hairless, bees with limited yellow maculations on the base of the tibiae and usually on the face. The face of the male is often extensively marked with ivory-white or yellow.

Eleven of the twelve species on the British list have been recorded from Kent. It is unclear if the twelfth, *Hylaeus pectoralis*, is a Kent species; there are no firm records although the species should be searched for at reed bed sites.

The species are usually aerial nesters, some species boring in the broken ends of bramble stems etc, whilst others will use dead wood containing beetle exit burrows. There may be an evolutionary process selecting for smaller size to fit such nest sites. *Hylaeus hyalinatus* frequently nests in soft mortar in old walls and *H. signatus* is found additionally in hard clay banks.

Hylaeus visit flowers with shallow corollas because of their short tongues and hence are frequent on Apiaceae. Brambles are another favoured host plant. The female carries pollen to the nest in the crop.

Hylaeus are the usual hosts of *Gasteruption* species (Hymenoptera: Gasteruptiidae).

Hylaeus annularis (Kirby)

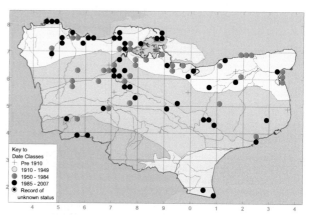

A common bee most frequently found on sandy soils in the north of the county, although there are a number of records from the chalk. The females visit *Rubus* (Rosaceae), Apiaceae and Asteraceae.

Occurrence: 47, 87   vi-ix   1897-2007
No status.

Hylaeus brevicornis Nylander

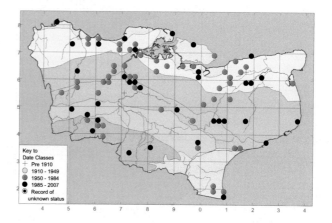

Fairly common and found on most soil types in the county, but perhaps scarcer on the weald clay. Nests in pithy stems and frequently reared from those of *Rubus* by Dr G H L Dicker. The females visit Apiaceae, *Rubus* (Rosaceae) and occasionally Asteraceae. Males have been found on *Sedum* (*pers. obs.*).

Occurrence: 34, 86 *vi-ix* 1898-2006
No status.

Hylaeus communis Nylander

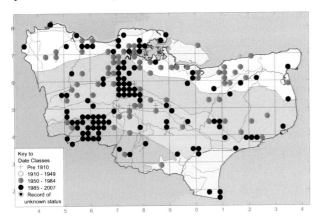

An abundant species found on most soil types in the county. The nest is constructed in pithy stems and beetle borings in dead wood. The female visits Asteraceae such as *Picris* and *Solidago*, Apiaceae and other flowers with shallow corollas.

Occurrence: 110, 189 *v-ix* 1896-2007
No status.

Hylaeus confusus Nylander

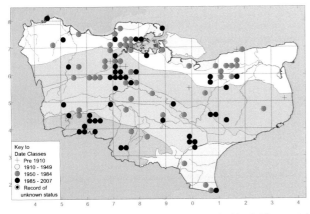

A common species found on most soil types in Kent. The nest is constructed in pithy stems and the female visits *Rubus* (Rosaceae) and Apiaceae.

Occurrence: 46, 101 *v-ix* 1894-2007
No status.

Hylaeus cornutus Curtis

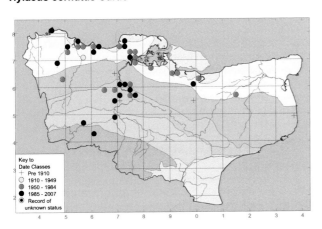

A local and rather scarce species. Most frequently recorded on the Eocene sands in the county and also from the Lower greensand in the Maidstone area. Apparently nests in pithy stems and galls but not reared by Dr G H L Dicker. The female forages on Apiaceae, Asteraceae and possibly *Reseda*. The pollen was once thought to be carried to the nest in a depression on the face of the female but this is now not regarded as likely.

Occurrence: 19, 43 *vi-viii* 1893-2006
National status Shirt: RDB3 Falk: pNa
Kent status Waite: Notable Present work: pKb.

Hylaeus gibbus Saunders

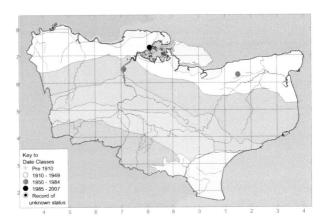

Usually a species of open heath and very rare in Kent as a result. Nests are likely to be in burrows in dead wood although undescribed for Britain. Also, pollen is likely to be collected from flowers with a shallow corolla, such as Apiaceae and *Rubus* (Rosaceae).

Occurrence: 1, 3 vii 1971-1987
National status Shirt: RDB3 Falk: pRDB3
Kent status Waite: KRDB1 Present work: pKRDB1

Hylaeus hyalinatus Smith

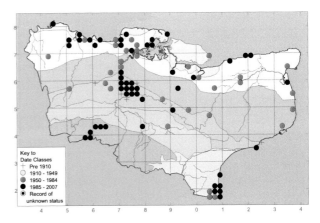

A common species most frequently recorded on sandy soils, also from coastal dunes and shingle. Scarce on the chalk and clay. Not reared from pithy stems by Dr G H L Dicker and it appears that the species prefers to nest in the soft mortar of old walls, possibly sometimes also in beetle exit holes in wood. The females are frequently found on umbellifers (Apiaceae) but as pollen is carried in the crop, samples are not readily obtained for analysis. Both males and females have been captured on *Hebe* and *Sedum* (*pers. obs.*). A putative host of the hymenopterous cleptoparasite, *Gasteruption minutum* (Gasteruptiidae).

Occurrence: 51, 88 *v-ix* 1878-2005
No status.

Hylaeus pictipes Nylander

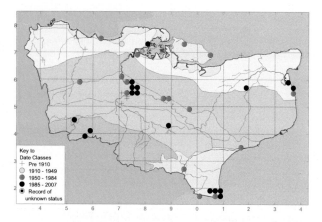

Recorded rather frequently but locally in the Maidstone and Dungeness areas, otherwise very scarce across the county. The female nests in beetle holes in old wood and in pithy stems. Reared by Dr G H L Dicker from *Anobium* borings in the dead stems of *Ulex europaeus*. Pollen sources are unknown in Britain but the bees have been found on Apiaceae, thistles (Asteraceae), *Reseda* and Brassicaceae. In Maidstone, a frequent visitor to *Hebe* (*pers. obs.*).

Occurrence: 17, 40   v-ix   1896-2005
National status   Falk: pNa
Kent status   Waite: Notable   Present work: pKb

Hylaeus punctulatissimus Smith

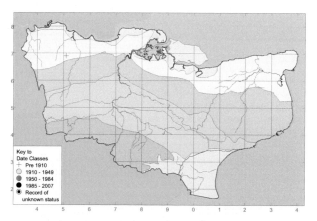

There are only two recorded British occurrences, one in Kent from Birch Wood, which no longer exists (coll. F Smith). The species is assumed extinct in the county and the UK, if indeed ever native.

Occurrence: 0, 1   A Kent flight period cannot be defined   1840
National status   Shirt: Appendix   Falk: Appendix
Kent status   Present work: pKRDB1+

Hylaeus signatus (Panzer)

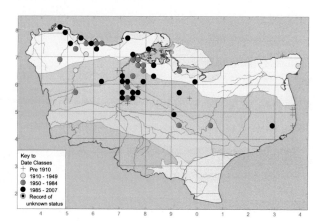

Well recorded from the East Thames corridor and the Medway valley but rather scarce elsewhere and apparently absent in the county south of the Lower Greensand. Nests most frequently in hard clay banks and in the mortar of old walls; occasionally in woody *Rubus* and *Rosa* stems. The only known pollen source is *Reseda* but the bees can also be found on flowers of Brassicaceae and Asteraceae. The males will additionally visit *Rubus* spp.

Occurrence: 25, 52   v-ix   1850s-2007
National status   Falk: pNb
Kent status   Waite: Notable   Present work: No Kent status.

Hylaeus spilotus Foerster
Synonymy: *H. euryscapus, H. masoni*

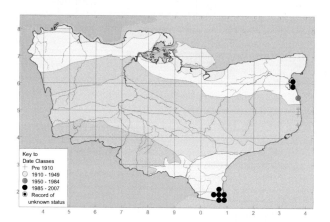

A rare bee recorded only on coastal sands and shingle in the county. The nesting requirements are unclear; the bee nests in pithy stems but the only British record is from loose sand. Pollen sources are unknown in Britain but the bees have been found on Apiaceae, Asteraceae and *Euphorbia*. At Sandwich and Greatstone-on-Sea both sexes of the bee have been found on *Sedum* (*pers. obs.*).

Occurrence: 8, 13   vi-viii   1882-2004
National status   Shirt: RDB3   Falk: pRDB3
Kent status   Waite: KRDB3   Present work: pKRDB3

Subfamily Andreninae

This subfamily is divisible into two British tribes, Andrenini and Panurgini, each with one indigenous genus. Both genera, *Andrena* and *Panurgus* respectively, are mining bees, i.e. they nest in burrows in the ground. They are united by a structural character: on the face of the bee there are two grooves rather than one running from each antennal socket down to the clypeus. This is unique to the subfamily.

In *Andrena* there are three submarginal cells in the forewing, whilst in *Panurgus* there are two. Additionally, the *Andrena* female has a felty inner ocular fovea (running down the inner margin of each compound eye). This is not felty in the male or in either sex of *Panurgus*.

Tribe Andrenini.

♀ *Andrena thoracica*

The only British genus is *Andrena*. World wide this is one of the largest animal genera and the species exhibit a wide range in size and general appearance. There are approximately 68 British species, of which 56 occur in Kent. A number of these are difficult to identify, particularly in the *A. minutula* complex. In the treatment given here, *A. nigrospina* is included under *A. pilipes*, although these are almost certainly specifically distinct.

The species are solitary apart from three in the *A. trimmerana* group. These are communal, sharing an entrance burrow but each female is believed to build off this her own nest burrow.

Andrena is the usual host of the cleptoparasitic bee genus *Nomada* (Apinae) and occasionally of *Sphecodes*, e.g. *S. pellucidus* (Halictinae).

Andrena alfkenella Perkins
Synonymy: *A. moricella*

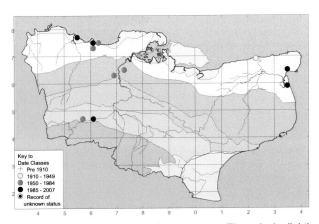

A rare species with two generations per year. These look slightly different in appearance and were once regarded as different species, hence the synonymy. Frequently coastal, both on soft rock cliffs and dunes, in the county but also from calcareous grassland and scarp faces of various strata. Nests in open areas in a variety of habitats but probably not in aggregations. Pollen sources are unknown in Britain. The first brood visits Brassicaceae, *Prunus spinosa* (Rosaceae), *Veronica* (Scrophulariaceae) and *Bellis* (Asteraceae); the second, Apiaceae, Brassicaceae, Rosaceae and *Veronica*.

Occurrence: 5, 12   *iii-vi, vii-viii*   1902-2005
National status   Shirt: RDB3   Falk: pRDB3
Kent status   Waite: KRDB1   Present work: pKRDB2

Andrena angustior (Kirby)

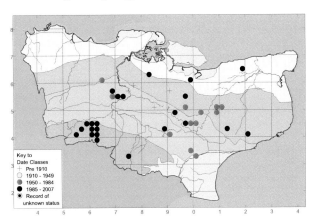

A widespread but rather uncommon species. The distribution is strongly and positively correlated with significant woodland in the county, particularly on chalk and the sands. Nesting is probably rather solitary and not in aggregations. Pollen sources are unclear but the females visit *Euphorbia amygdaloides*, *Salix* and *Ranunculus*. The male has been found on *Taraxacum*.

Occurrence: 23, 39   *iv-vi*   1879-2005
No status.

Andrena apicata Smith

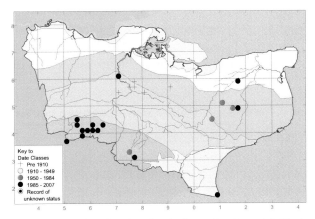

A scarce mining bee found on the chalk, sand and coastal shingle. Nesting is in light soils in warm situations, such as banks and sometimes level ground. The females forage almost exclusively on *Salix* catkins but occasionally on *Ulex*. A host of *Nomada leucophthalma*.

Occurrence: 15, 26   *iii-v*   1893-2007
National status   Falk: pNb
Kent status   Waite: Notable   Present work: pKb

Andrena barbilabris (Kirby)

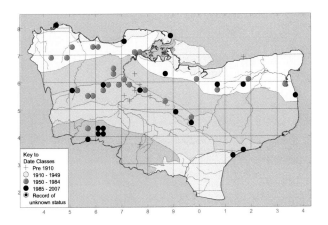

A widespread but uncommon species associated with sandy soils, occasionally coastal. The flight period is exceptionally long, there being only one generation per year. The female nests in sandy soils, sometimes in very loose sand. The bees have been recorded from *Vaccinium*, *Taraxacum* and *Smyrnium*, though it is unclear if these are pollen sources. The host of *Sphecodes pellucidus*.

Occurrence: 19, 52   *iii-vii*   1892-2007
No status.

Andrena bicolor Fabricius

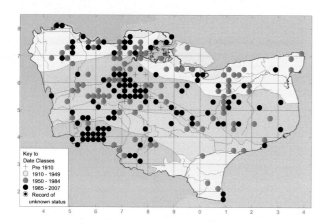

An abundant species in the county, with two generations per year. The phenology shows two strong modes slightly overlapping. The species is frequent on chalk and the sands, less so on clay soils. The female nests in light soils in warm situations. Pollen sources are *Taraxacum*, *Tussilago*, *Ranunculus*, Brassicaceae, *Salix*, *Prunus spinosa*, *Rubus*, *Mentha*, *Viola*, *Fragaria*, *Smyrnium*, *Narcissus* and further genera for the second generation. A host of *Nomada fabriciana*.

Occurrence: 107, 210   *ii-v, vi-viii*   1892-2007
No status.

Andrena bimaculata (Kirby)

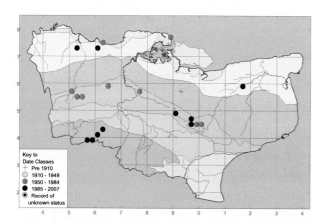

A scarce species with two generations per year. It is strongly associated with sandy soils, rarely coastal. Nests in dry, sandy soils and pollen sources include *Salix*, Asteraceae and Rosaceae for the first generation. The second forages on *Rubus*, Apiaceae and other flowers. A host of the cleptoparasitic bee, *Nomada fulvicornis*.

Occurrence: 10, 20   *ii-v, vi-viii*   1890s-2007
National status   Falk: pNb
Kent status   Waite: Notable   Present work: pKa

Andrena bucephala Stephens

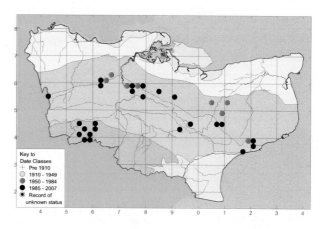

I hesitate to give this mining bee county scarcity status, as it seems to be too widely recorded for this. Most often recorded in May. A species often found on the chalk scarps but also on sandy soils. The females nest communally in soil which can be very calcareous and forage on tree blossom, mostly *Crataegus* but occasionally *Acer* and others. Host to *Nomada hirtipes* and probably some other *Nomada* species.

Occurrence: 24, 32   *iii-vi*   1966-2004
National status   Shirt: RDB3   Falk: pNa
Kent status   Waite: Notable   Present work: No Kent status

Andrena carantonica Perez
Synonymy: *A. scotica*, *A. jacobi*. Confused in the past as *A. trimmerana* and *A. rosae*.

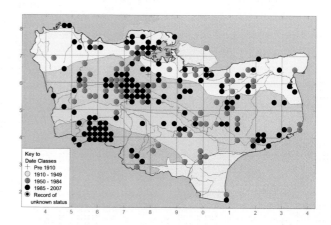

This species has been much confused in the past under other names. It is an abundant bee but not frequent on the weald clay. Nesting takes place in light soils. The female forages mainly on tree blossom, including that of *Pyrus*, *Malus* and *Crataegus* (Rosaceae). Host of the cleptoparasitic bees, *Nomada flava*, *N. marshamella* and possibly other species of *Nomada*.

Occurrence: 128, 208   *iii-viii*   1893-2007
No status.

Andreninae 65

♀ *Andrena carantonica*
© Steve Smith

Recorded again in the county recently after a gap of many decades. This is consistent with the recorded expansion of the species in other southern counties. Mainly from sandy soils. I have no information on flower visits. The host of *Nomada lathburiana*.

Occurrence: 22, 24   *iii-v, vii*   1896-2007
No status.

Andrena chrysosceles (Kirby)

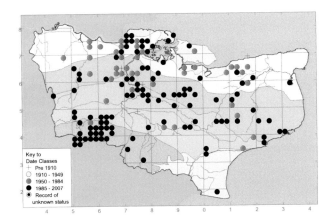

An abundant bee most frequent on the sands and chalk but not scarce on clays. The late records may be a partial second brood. A visitor to various plants such as *Ranunculus, Euphorbia, Heracleum, Berberis, Prunus, Salix, Taraxacum, Tussilago, Crataegus, Rosa, Pyracantha* and Brassicaceae. Many of these will represent pollen sources. Both sexes are frequently stylopised.

Occurrence: 106, 151   *iii-vii, x*   1893-2007
No status.

Andrena cineraria (Linnaeus)

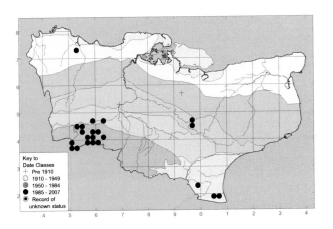

Andrena clarkella (Kirby)

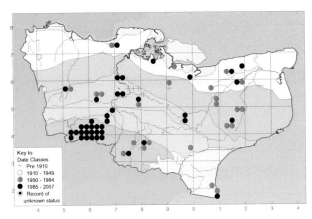

A common bee from the sands and chalk, but mainly recorded from the High Weald. A very early mining bee usually over by early May; the female visits *Salix* almost exclusively. Host to *Nomada leucophthalma*.

Occurrence: 41, 72   *ii-v*   1891-2007
No status.

Andrena coitana (Kirby)

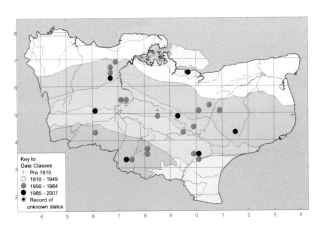

Although well known from the county, this species is showing a strong and significant decline, there being only two post-1991 records known for Kent. A late flying species. It shows a positive correlation with significant woodland but not specific to any particular soil type. The females visit mayweeds, thistles, *Euphorbia amygdaloides, Rubus* and *Heracleum*. Host of the cleptoparasitic bee, *Nomada obtusifrons*, not known from the county.

Occurrence: 7, 29   *vi-viii*   1894-2002
Kent status   Present work: pKa

The Bees, Wasps and Ants of Kent

Andrena denticulata (Kirby)

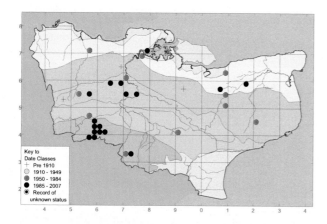

Rather local in the county and another very late flying species. Mainly from woodland on sandy soils. The females visit Asteraceae including *Pulicaria dysenterica*, Apiaceae including *Torilis*, and *Trifolium* (Fabaceae).

Occurrence: 17, 31   *vii-ix*   1890s-2007
No status.

Andrena dorsata (Kirby)

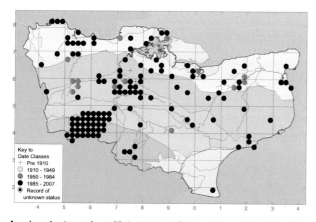

An abundant species with two generations per year. The phenology shows two strong and largely separated modes. The species may have increased in recent decades. Mainly recorded on the sand with some occurrences on chalk; scarce on the clay and rarely coastal. The species is widely polylectic on Rosaceae, Asteraceae, Apiaceae etc.

Occurrence: 122, 139   *iii-vi, vi-ix*   1896-2007
No status.

Andrena falsifica Perkins

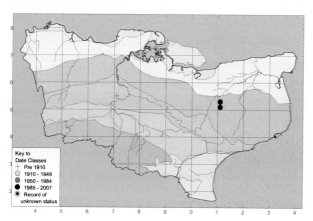

This appears to be a genuinely rare bee in the county and has not been recorded recently. There are too few data to judge soil type preferences but heath and calcareous grassland are mentioned in the literature (Falk, 1991). Nesting preferences are unknown and pollen sources have yet to be established in Britain. The species has been captured on *Bellis* and *Taraxacum* (Asteraceae) in the county.

Occurrence: 2, 3   *iv-vi*   1896-1985
National status   Shirt: RDB3   Falk: pNa
Kent status   Waite: Notable   Present work: pKRDB2

Andrena ferox Smith

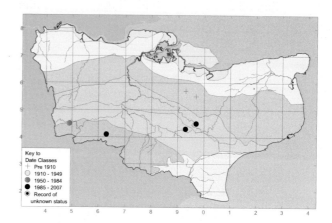

A very rare bee in the county, a recognised stronghold for the species in Britain. It has been recorded only sporadically over the years. A communal nester, sometimes with many females sharing a common burrow entrance. They forage on tree blossom, including that of *Quercus* and *Crataegus*. A probable host of *Nomada marshamella* and *N. flava*.

Occurrence: 3, 6   *v-vi*   1896-2002
National status   Shirt: RDB1   Falk: pRDB1
Kent status   Waite: KRDB1   Present work: pKRDB1

Andrena flavipes Panzer

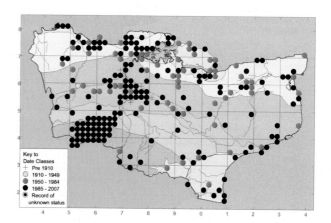

This is an abundant bee in Kent, with two generations per year. There is only slight overlap between the two broods. Rather scarce on the weald clay but otherwise found on all soil types; however most frequent on sandy soil. The female is widely polylectic, foraging on *Taraxacum* and other Asteraceae, *Rubus* (Rosaceae), Brassicaceae, *Smyrnium*, *Salix* and many other plants. The host of *Nomada fucata*, also a species with two generations per year.

Occurrence: 167, 235   *ii-vi, vi-xi*   1871-2007
No status.

Andreninae

Andrena fucata Smith

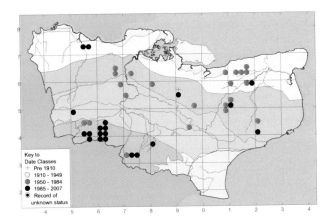

A widespread species but not frequent. Found particularly on the chalk and the High Weald sands. There may be two broadly overlapping modes in frequency but this may be due to observer bias rather than it being a double brooded species; it has a long flight period. The female has been recorded from *Euphorbia*, *Ranunculus*, *Rubus* and *Ballota* but which if any of these are pollen sources is unclear.

Occurrence: 20, 44   iv-viii   1890s-2004
No status.

♀ *Andrena fulva*
(the female *A. fulva* is sometimes misidentified as the worker of *Bombus pascuorum*)
© Lee Manning

Andrena fulva (Mueller in Allioni)
Synonymy: *A. armata*

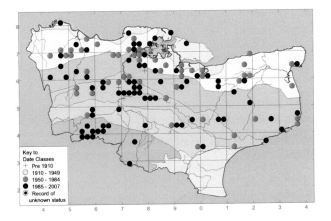

A very common species often recorded digging burrows in garden lawns. Most frequent on sandy soils. Widely polylectic; forage plants include *Taraxacum*, *Prunus*, *Salix*, *Euphorbia*, *Vaccinium*, *Ulex* and others. The host of the rare cleptoparasitic bee, *Nomada signata*.

Occurrence: 75, 130   iii-vii   1893-2007
No status.

Andrena fulvago (Christ)

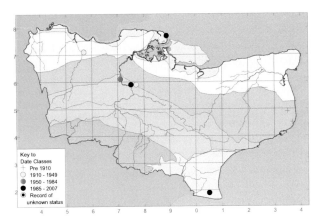

A rare bee but one with a long history in the county. Most modern Kent records are estuarine or from coastal shingle. Often recorded in June. Females have been recorded from yellow flowered Asteraceae such as *Hypochaeris* and *Taraxacum*.

Occurrence: 3, 9   iv-viii   1897-2007
National status   Shirt: RDB3   Falk: pNa
Kent status   Waite: Notable   Present work: pKRDB2

Andrena fuscipes (Kirby)

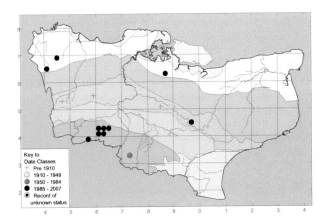

Andrena fuscipes is very much a species of sandy heath and so is scarce in the county. Most records are from August when Ericaceae are in bloom, the exclusive forage plant family. A host of *Nomada rufipes*.

Occurrence: 10, 14   vii-ix   1895-2007
Kent status   Present work: pKa

Andrena gravida Imhoff

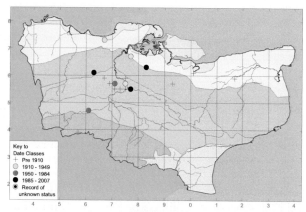

Although this bee was fairly frequent in the Maidstone area during the late 19th and early 20th centuries, it has since declined drastically and is recorded only sporadically now, being genuinely rare. The main flight period is March to May but with a possible small, summer emergence. The female forages on *Malus*, *Prunus*, *Salix* and *Taraxacum*, and possibly other plants.

Occurrence: 3, 18   *iii-v, vii-viii*   1880-2006
National status   Shirt: RDB1   Falk: pRDB1
Kent status   Waite: KRDB1   Present work: pKRDB2

Andrena haemorrhoa (Fabricius)

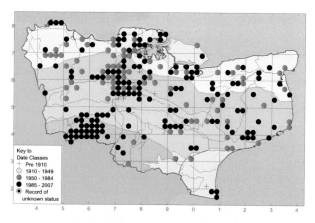

An abundant bee in the county recorded mainly on sand and chalk. The main flight period is April to May and the phenology curve is unimodal. A widely polylectic spcies, foraging from plants as varied as *Prunus*, *Crataegus*, *Euphorbia* and *Taraxacum*. Host of the cleptoparasitic bee, *Nomada ruficornis*.

Occurrence: 150, 234   *ii-viii*   1893-2007
No status.

Andrena hattorfiana (Fabricius)

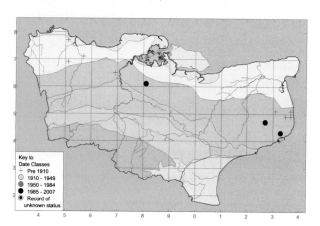

A large, conspicuous mining bee unlikely to be much under recorded, hence considered extremely rare. Found exclusively on calcareous grassland in the county. It has a late flight period, timed to coincide with the flowering of its host plant, *Knautia arvensis*, field scabious. Host of the rare cleptoparasitic bee, *Nomada armata*.

Occurrence: 3, 11   *vii-viii*   1850s-2007
National status   Shirt: RDB2   Falk: pRDB3
Kent status   Waite: KRDB2   Present work: pKRDB2

Andrena helvola (Linnaeus)

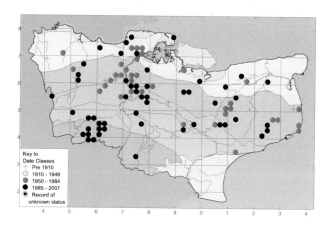

Common; mainly from the sands and chalk, where it can be numerous. Although the flight period is quite long, it is also unimodal; therefore the bee has only one generation per year. Forages widely, on plants such as Brassicaceae, Apiaceae, Asteraceae and others. A putative host of *Nomada panzeri*.

Occurrence: 52, 95   *iii-vi*   1894-2006
No status.

Andrena humilis Imhoff

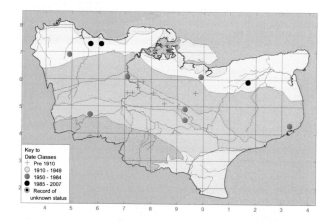

A very scarce species, mainly from sandy soils. The recorded flight period in Kent is quite short, from 13 May to 20 June, allowing a comparatively small window of opportunity in which to collect it. The nest is excavated in hard sand, either level, sloping or in banks. The females appear to collect pollen exclusively from yellow flowered Asteraceae such as *Pilosella*, *Picris*, *Crepis* and *Taraxacum* and their activity is mainly confined to the morning, when *Picris* flowers are open. Host of the scarce cleptoparasitic bee, *Nomada integra*.

Occurrence: 3, 19   *v-vi*   1850s-1999
National status   Falk: pNb
Kent status   Waite: Notable   Present work: pKRDB3

Andrena labialis (Kirby)

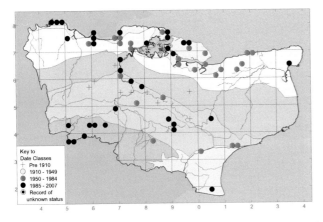

A fairly common bee found on most soils types, including coastal soft rock cliffs but rarely recorded on the chalk. Most records are from June and the phenology curve is unimodal. Females are to be found foraging on various flowers, including *Cytisus*, *Medicago*, *Trifolium* and *Euphorbia*. The male is also found on *Ranunculus* and Dr G H L Dicker found a specimen with orchid pollinia attached. Host of the cleptoparasitic bee, *Sphecodes rubicundus*.

Occurrence: 33, 69   *v-viii*   1897-2006
No status.

Andrena labiata Fabricius

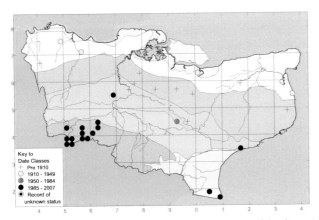

The recorded flight period is early April to mid-June and the August date requires validation. This is a scarce species with modern data only on the sands and coastal shingle. There are old records on the chalk. It is to be found in a variety of habitats on the sand but appears to forage mainly on *Veronica chamaedrys* (Falk, 1991). Host of the rare cleptoparasitic bee, *Nomada guttulata*.

Occurrence: 16, 32   *iv-vi*, [*viii*]   1878-2005
National status   Shirt: RDB3   Falk: pNa
Kent status   Waite: Notable   Present work: pKb

Andrena lapponica Zetterstedt

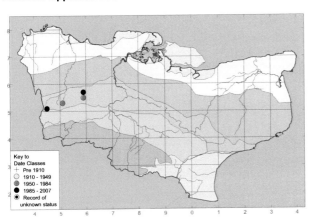

A rare species in the county, recorded only on the Lower greensand in the west. A heath and moorland species, apparently common in some other counties, often in the north and west of the UK. Most Kent records are from April and early May, the June record being unusual. The females appear to forage only on *Vaccinium* flowers but will take nectar from *Taraxacum*.

Occurrence: 2, 4   *iv-vi*   1895-2005
Kent status   Present work: pKRDB3

Andrena marginata Fabricius

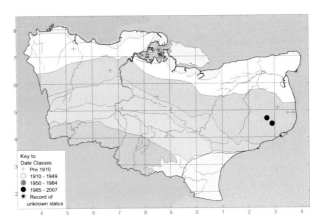

A rare bee but with a long history in the county. Modern data are only from the Lydden/Temple Ewell area. Recorded solely from chalk grassland, where the female obtains pollen from *Scabiosa* and *Succisa*. Flower visits to other species are probably only for nectar. Host of the rare cleptoparasitic bee, *Nomada argentata*.

Occurrence: 2, 8   *vii-viii*   1850s-2002
National status   Falk: pNa
Kent status   Waite: Notable   Present work: pKRDB2

Andrena minutula (Kirby)
Synonymy: *A. parvula*

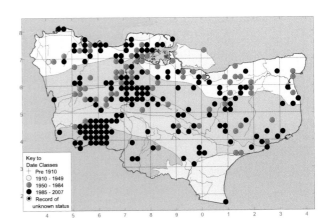

An abundant bee in the county and found widely, although scarcer on the weald clay. The phenology is clearly bimodal, indicating two generations per year. These are sufficiently different in the male for them to be once regarded as different species. The female is widely polylectic, visiting such diverse plant families as Ranunculaceae, Salicaceae, Asteraceae, Apiaceae, Rosaceae and Brassicaceae. Host of the cleptoparasitic bees, *Nomada flavoguttata* and a small form of *N. fabriciana*.

Occurrence: 158, 230   *iii-vi*, *vi-ix* [*x*]   1893-2007
No status.

Andrena minutuloides Perkins
Synonymy: *A. parvuloides*

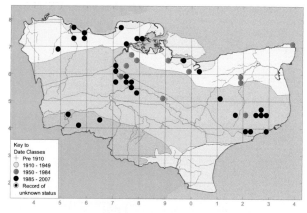

There are too many modern records for this to be regarded here as a notable species, although confusion with the last must occur. Found on the sands and particularly the chalk, but is rare on clays. Another species with two generations per year, the second being more frequent. Nesting probably takes place in light soils which are short-cropped or sparsely vegetated. The first brood visits various plant families, including Brassicaceae, Rosaceae and Asteraceae, whilst the second seems to favour Apiaceae.

Occurrence: 31, 42   *iv-vi, vii-ix*   1957-2006
National status   Falk: pNa
Kent status   Waite: Notable   Present work: No Kent status.

Andrena nana (Kirby)

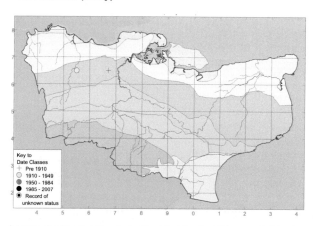

Assumed extinct not only in Kent but in the UK as well. The two known Kent occurrences were on the chalk in West Kent (Luddesdown, coll. H Elgar; Eynsford, coll. G E J Nixon). The bee has two generations per year. Pollen sources are unknown in the UK.

Occurrence: 0, 2   *iv, viii*   1899, 1930
National status   Shirt: RDB1   Falk: pRDB1+
Kent status   Present work: pKRDB1+

Andrena nigriceps (Kirby)

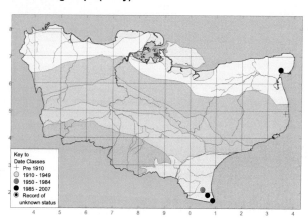

A rare bee in the county and only from coastal areas. Nesting is likely to occur in light soil. The female has been recorded as taking pollen from *Filipendula, Rubus, Castanea, Jasione, Galium* and Asteraceae in the UK.

Occurrence: 3, 5   *vi-vii*   1861-2007
National status   Falk: pNb
Kent status   Waite: Notable   Present work: pKRDB3

Andrena nigroaenea (Kirby)

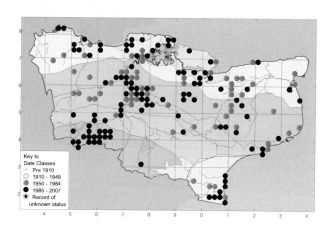

This is an abundant species in the county, less frequent on the weald clay. Although this bee has a long flight period, the phenology is unimodal, suggesting only one generation per year. The female is widely polylectic, foraging on *Taraxacum, Salix, Smyrnium, Euphorbia, Malus* and diverse other plants. Males have also been recorded from *Prunus spinosa*. Host to the cleptoparasitic bee, *Nomada goodeniana*.

Occurrence: 105, 172   *iii-ix*   1893-2007
No status.

♀ *Andrena nigroaenea*
© Jeremy Early

Andrena nitida (Mueller)
Synonymy: *A. pubescens*

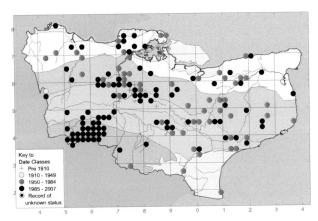

Also an abundant species but not quite as frequent as the last. From most soil types but less frequent on the weald clay. With a long, unimodal, flight period. The female is widely polylectic, although *Smyrnium olustratum* is often mentioned as being visited. Also frequently visited are *Taraxacum* and *Prunus spinosa*. Host to the cleptoparasitic bee, *Nomada goodeniana*.

Occurrence: 90, 149   *iii-vii*   1893-2007
No status.

Andrena niveata Freise
Synonymy: *A. spreta*

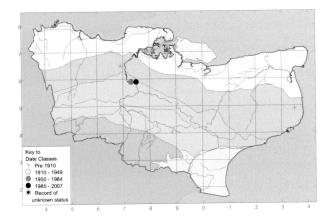

A very rare species but recently recorded also from other south-east counties: Surrey (D W Baldock, *pers. comm.*) and East Sussex (M Jenner, *pers. comm.*). Modern Kent records appear to be from sandy soil close to the gault clay but it formerly occurred more widely. Pollen is collected mostly if not entirely from Brassicaceae.

Occurrence: 1, 6   *v-vi*   1897-2005
National status   Shirt: RDB3   Falk: pRDB2
Kent status   Waite: KRDB1   Present work: pKRDB1

Andrena ovatula (Kirby)

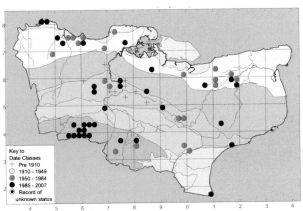

A common species recorded predominantly on sandy soils in the county. There is a possible decline in the Maidstone area. The phenology indicates a life cycle with two broods, although the two modes overlap sufficiently that some specimens cannot be assigned with confidence to their correct generation. The female forages on a range of plants, including *Taraxacum*, *Euphorbia*, *Erica* and *Cytisus*.

Occurrence: 33, 63   *iv-vi, vi-ix*   1879-2007
No status.

Andrena pilipes aggregate
Including *A. pilipes* (*sens. str.*) and *A. nigrospina*.

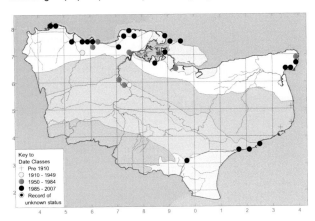

Unfortunately, these two species can only be separated by microscopic examination of males, which have not been available for study. The characters of the female are subtle and subject to variation, hence not reliable. Most coastal records will apply to *A. pilipes* (*sens. str.*) whilst many inland will refer to *A. nigrospina*. *A. pilipes* has two generations, while *A. nigrospina* has only one. Both species forage on a range of plants, including Brassicaceae and *Cytisus;* males have been recorded from *Prunus spinosa*. Both species may be hosts to the cleptoparasitic bee, *Nomada fulvicornis*.

Occurrence: 22, 35   *iv-ix*   1850s-2007
National status   Falk: pNb
Kent status   Present work: No Kent status.

Andrena polita Smith

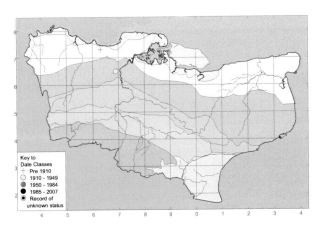

Presumed long extinct in the county and country. It had been recorded in the UK only on the chalk in West Kent. Smith's 1847 specimens came from chalk pits at Northfleet and several recorders took the species at Cuxton Warren, Upper Halling in the early part of the 20th century. My understanding is that all known British specimens were captured in July, although Falk states May. British pollen sources are unknown. No *Nomada* species were known on this bee.

Occurrence: 0, 2   *vii*   1847-1934
National status   Shirt: RDB1+   Falk: pRDB1+
Kent status   Present work: pKRDB1+

72 The Bees, Wasps and Ants of Kent

Andrena praecox (Scopoli)

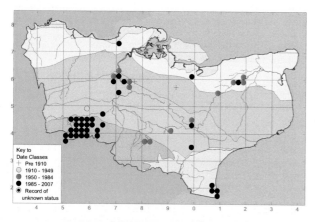

A rather local species found predominantly on the sands and coastal shingle, although there are some data for the weald clay. The June record is exceptionally late; by mid-May specimens are usually very worn and sun bleached. An early species that forages predominantly on *Salix* catkins. Host to the rare cleptoparasitic bee, *Nomada ferruginata* (= *xanthosticta*).

Occurrence: 36, 51   *iii-vi*   1880-2007
No status.

Andrena proxima (Kirby)

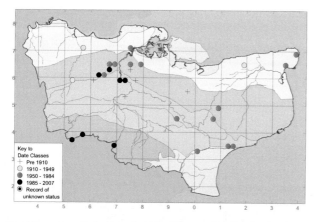

Kent is apparently a stronghold for this species. Particularly found on the chalk although present on scarp faces of other strata. Nesting occurs in light soils, which are sparsely vegetated or have short turf. The female forages particularly on Apiaceae. Host to the rare cleptoparasitic bee, *Nomada conjungens*.

Occurrence: 7, 29   *iv-vii*   1898-2006
National status   Shirt: RDB3   Falk: pRDB3
Kent status   Waite: KRDB3   Present work: pKa

Andrena rosae Panzer

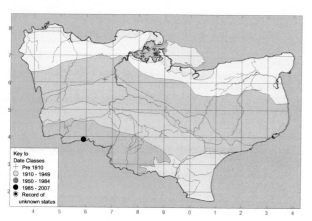

A very rare species in the county that may have declined nationally. The statuses need to be re-assessed following the separation of the spring brood as a distinct species, *A. stragulata*. Nesting is poorly known. The true *A. rosae* is known to forage on Asteraceae, Apiaceae and *Rubus* (Rosaceae).

Occurrence: 1, 2   *vii*   1896, 2000
National status   Shirt: RDB3   Falk: pRDB2
Kent status   Waite: KRDB1   Present work: pKRDB1

Andrena semilaevis Perez
Synonymy: *A. saundersella*

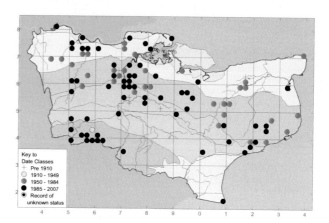

A common bee particularly frequent on the chalk. The females particularly visit Apiaceae for pollen, other flower records may possibly be for nectar only. A probable host of the cleptoparasitic bee, *Nomada flavoguttata*.

Occurrence: 61, 100   *iv-viii*   1897-2007
No status.

Andrena similis Smith
Synonymy: *A. ocreata*

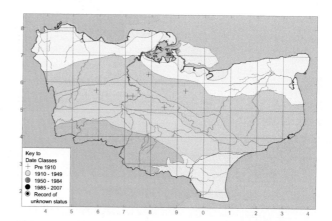

A species apparently locally frequent in the Maidstone area around 1900 but drastically declined in the county soon afterwards. May now be extinct in Kent but was recorded in Sussex as recently as 1985. Flower visits to *Salix* and *Vaccinium* have been recorded but it is unclear if these were for nectar only.

Occurrence: 0, 7   *iv-v*   1897-1902
Kent status   Present work: pKRDB1

Andreninae 73

Andrena simillima Smith

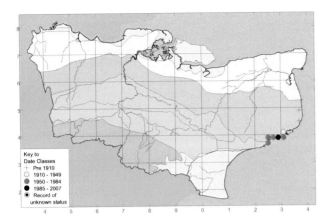

A rare species found in Kent only along the coast from Folkestone to Kingsdown. There may have been a slight contraction of range but this could also be a result of under recording. It is important to confirm the continued presence of the species in the county, as the last known record was in 1989. Nesting habits and pollen preferences are unclear.

Occurrence: 1, 8 *vii-viii* 1850s-1989
National status Shirt: RDB3 Falk: pRDB2
Kent status Waite: KRDB2 Present work: pKRDB2

Andrena stragulata Illiger
Synonymy: *A. eximia*

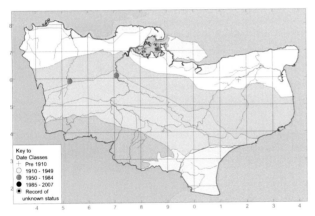

Formerly regarded as the spring generation of *A. rosae*, now assumed a distinct species. A very rare bee which may prefer hot spots such as scarp faces. The species is recorded from *Prunus spinosa* and *Salix* but it is unclear if the former represents only a nectar source.

Occurrence: 0, 3 *iv* 1880-1978
Kent status Present work: pKRDB1

Andrena subopaca Nylander

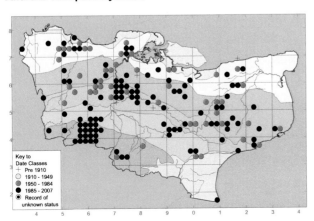

An abundant bee with an exceptionally long flight period. The phenology curve is essentially unimodal, indicating only one generation a year. Prevalent on sand and chalk soils, and scarce on the weald clay. Often a woodland species. The species is widely polylectic, visiting plants in the families Ranunculaceae, Asteraceae, Brassicaceae and others. A possible host of the cleptoparasitic bee, *Nomada flavoguttata*.

Occurrence: 105, 158   *iii-ix*   1892-2007
No status.

Andrena synadelpha Perkins

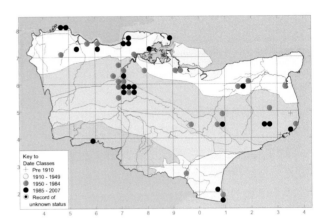

A rather local bee frequently recorded from the Aylesford area and north Kent, otherwise with only scattered records. Virtually absent from clay soils. The phenology curve is unimodal, the July record being exceptionally late. The female visits *Crataegus*, *Prunus* and *Acer pseudoplatanus*, amongst other species.

Occurrence: 23, 50 *iv-vii* 1897-2005
No status.

Andrena tarsata Nylander

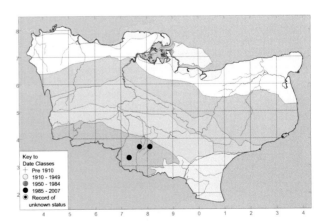

Only known in the county from the collecting of Dr G H L Dicker and a nationally declining species. Found very locally in the Cranbrook/ Benenden/ Bedgebury area. The species is known to be oligolectic on *Potentilla*. Host of the rare cleptoparasitic bee, *Nomada roberjeotiana*, which is unknown from the county.

Occurrence: 3, 3   *vii-viii*   1979-1988
Kent status   Present work: pKRDB3

Andrena thoracica (Fabricius)

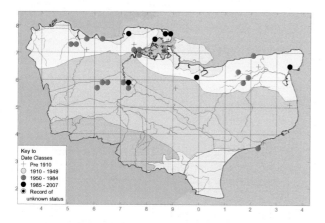

Not frequent in the county and apparently showing a decline. It prefers sandy soils and coastal soft rock cliffs – sand and gravel pits are often recorded as sites. A species with two generations per year. Flower visits, some undoubtedly for pollen, are to *Tussilago*, *Salix*, *Prunus*, *Euphorbia*, *Viola* and other plants. A possible host of the cleptoparasitic bee, *Nomada goodeniana*.

Occurrence: 7, 31   *iii-vi, vii-ix*   1890s-2004
Kent status   Present work: pKa

Andrena tibialis (Kirby)

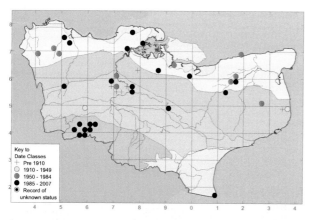

A scarce species recorded mainly from sandy soils, usually heathland. The bee is possibly showing a decline. It closely resembles some forms of *A. bimaculata*. The phenology curve is essentially unimodal, implying a single generation per year. The bee is polylectic although frequently recorded from *Salix*. A host of the cleptoparasitic bee, *Nomada fulvicornis*.

Occurrence: 23, 41   *iii-vi*   1891-2007
National status   Shirt: RDB3   Falk: pNa
Kent status   Waite: Notable   Present work: pKb

Andrena trimmerana (Kirby)
Synonymy: *A. spinigera*

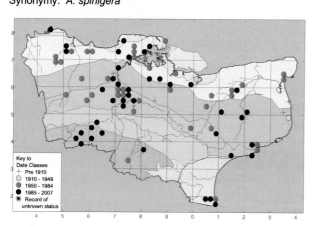

There has been much confusion in the past over the application of this name. The bee is relatively frequent in Kent and can be considered fairly common here, although scarce on the clay. It has two generations per year. It is possible that in the future these will be regarded as separate species. The first brood is the more frequent. Nesting takes place in sandy banks and in the earth of root plates of uprooted trees (*pers. obs.*). The spring brood is often recorded on *Salix* and *Prunus spinosa* and the second on *Rubus*, thistles and Apiaceae. A possible host of the cleptoparasitic bee, *Nomada marshamella*.

Occurrence: 41, 87   *iii-v, vi-ix*   1882-2007
National status   Falk: pNb
Kent status   Waite: Notable   Present work: No Kent status.

Andrena vaga Panzer

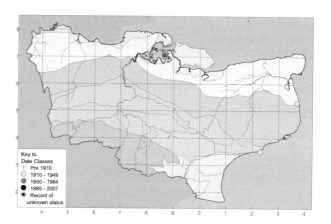

Probably extinct in the UK. Recorded in Kent from Deal (coll. K M Guichard) and Folkestone (coll. G B Collins). It is oligolectic on *Salix*.

Occurrence: 0, 2   *iv*   1939, 1946
National status   Shirt: RDB1   Falk: pRDB1+
Kent status   Waite: KRDBX   Present work: pKRDB1+

Andrena varians (Rossius)

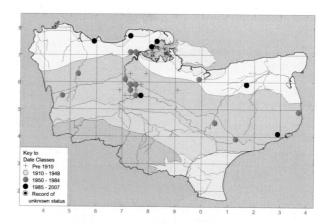

A scarce, apparently declining, bee. Not recorded south of the lower greensand in Kent. With an essentially unimodal phenology. The nest is dug in short grass, sometimes in semi-shade (*pers. obs.*). Apparently, there is an association with spring flowering fruit trees, such as *Prunus*, *Pyrus* and possibly *Malus*. A probable host of the cleptoparasitic bee, *Nomada panzeri*.

Occurrence: 7, 29   *iii-vi*   1894-1991
National status   Falk: pNb
Kent status   Waite: Notable   Present work: pKa

Andreninae to Halictinae

Andrena wilkella (Kirby)

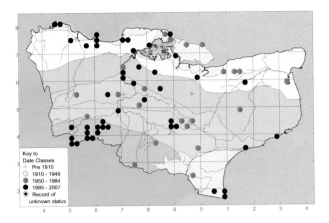

A common bee on the sands, both sedimentary and alluvial, and coastal shingle. Less frequent on the chalk but found on clay soils. At Dungeness the bee nests in banks of shell sand. With a unimodal phenology, most records during late May and June. Oligolectic on Fabaceae but will also visit *Crataegus* and *Rubus*, presumably for nectar only. Host of the cleptoparasitic bee, *Nomada striata*.

Occurrence: 45, 69 iv-viii 1894-2007
No status.

Tribe Panurgini.

♀ *Panurgus calcaratus*

The only British genus is *Panurgus*, with two indigenous species. Both occur in the county but neither is particularly common. However, *P. calcaratus* can be locally frequent. The species are known to favour yellow flowered Asteraceae as forage and the males can sometimes be found aestivating for short periods in them.

Panurgus banksianus (Kirby)

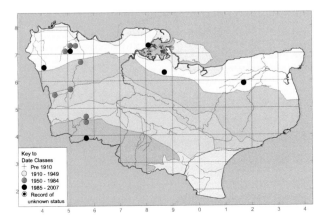

A scarce bee in Kent, mostly from sandy soils. Possibly under recorded in the county – most known records are fairly recent. Flower visits, some probably for pollen, are to yellow flowered Asteraceae such as *Hieracium*.

Occurrence: 6, 15 vi-viii 1890s-2007
Kent status Present work: pKa

Panurgus calcaratus (Scopoli)

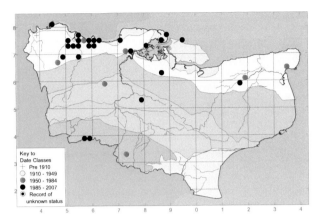

Rather local but apparently frequent on the marshes at Erith and Swanscombe. Scattered records elsewhere but only on sandy soils. Flower visits are to yellow flowered Asteraceae, including *Picris hieracioides* and *P. echioides*.

Occurrence: 25, 33 vi-ix 1850s-2007
No status.

Subfamily Halictinae

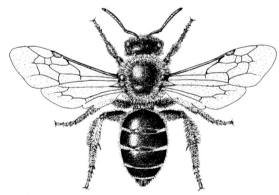

♀ *Halictus rubicundus*

The three extant British genera are assigned to the tribe Halictini. These are *Halictus*, *Lasioglossum* and *Sphecodes*. Kent has two common *Halictus* species, the remainder in the genus being rare or extinct. *Lasioglossum* is very speciose, with some 31 British species. Of these, 23 are firmly recorded from Kent. Two further *Lasioglossum* species may occur in the county: *L. cupromicans* has so far not been found here but is possibly increasing in other southern counties. There is a possible record for *L. prasinum* from TQ96 but I have no firm data. This latter species is described as oligolectic on Ericaceae although it must forage from other plants before the heathers are in flower.

Halictus and *Lasioglossum* are mining bees, the many smaller species sometimes being called sweat bees. They are notorious for their numerous species groups, the siblings of which can be very difficult to separate. Species of the *Sphecodes geoffrellus* group can also be difficult, particularly the females.

Halictus and *Lasioglossum* have a life cycle which predisposes them to sociality. The males emerge from mid summer to autumn and mate with the females, which then over winter. The solitary females emerge in the spring to found nests. These elements in the halictine cycle are to be found in social wasps and bumble bees. In some halictines the first reared brood is all female and these usually help the foundress to establish a primitive society. The complexity of halictine life cycles is still being unravelled and a range of social organisation types has been described in the literature. In a small number of species, the different female castes are to some extent morphologically defined. This includes the Kentish *Lasioglossum malachurum*, where the foundresses are larger and more robust than the all-female broods reared early in the colony cycle.

The 16 British species of *Sphecodes* have all been found in Kent, although there is a further Channel Islands species. They are cleptoparasites, mainly of *Lasioglossum* but also *Halictus* and *Andrena*. The species found on the primitively social *Lasioglossum* could justifiably be called social parasites, whilst those habitually on *Andrena* often have males which emerge in the spring with the females, as in that mining bee genus.

Tribe Halictini.

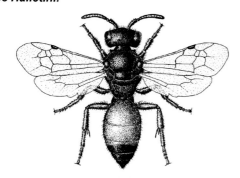

♀ *Sphecodes ephippius*

Halictus confusus Smith

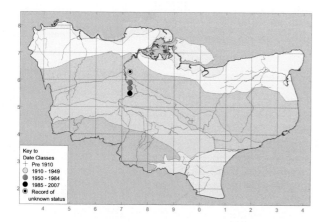

Unfortunately, the recording of this species in the county has been hampered by its very close similarity to the common *H. tumulorum*: the female, which is the sex more likely to be collected, is particularly close. Nonetheless, a female from Barming Heath in 2002 was confirmed by G R Else as *H. confusus*. Most of the other dots are likely to prove correct but that from Monk Wood, Burham, is on the chalk and more doubtful. The species is typical of sandy heath. This is a mining bee which may prove eusocial. It is probably widely polylectic, as in many other *Halictus*.

Occurrence: 1, 5   v-vii   1903-2002
National status   Shirt: RDB3   Falk: pRDB3
Kent status   Waite: KRDBK   Present work: pKRDB2

Halictus eurygnathus Bluethgen

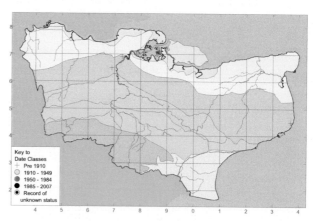

Apparently not infrequently encountered in the county around 1900 but from very few localities. Although assumed extinct in Kent, it was recently rediscovered in East Sussex, from several localities (S J Falk, *pers. comm.*). Hence, it should be searched for on coastal chalk in areas of rich floral diversity, particularly those with *Centaurea scabiosa*, its main forage plant, in abundance.

Occurrence: 0, 3   v-ix   1895-1907
National status   Shirt: RDB1+   Falk: pRDB1+
Kent status   Present work: pKRDB1+

Halictus maculatus Smith

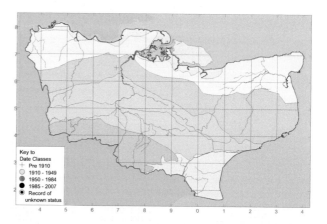

This was always a rare species and is assumed long extinct. It forages mainly on Asteraceae.

Occurrence: 0, 2   viii   1900
National status   Shirt: RDB1+   Falk: pRDB1+
Kent status   Present work: pKRDB1+

Halictus rubicundus (Christ)

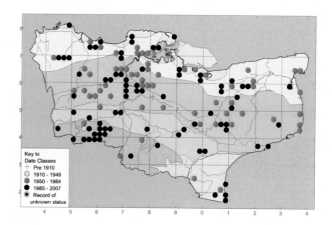

Halictinae 77

A very common species found predominantly on the sands and chalk, frequently coastal. The phenology curve is trimodal, indicating a eusocial species. The first mode is likely to be of spring foundress females, the second of workers and the third of autumn males and over-wintering females. Those records not coastal appear to be positively but less than strongly, correlated with established woodland. The female forages on Asteraceae and *Euphorbia*, and probably many other plant families. A host of the cleptoparasitic bee, *Sphecodes monilicornis*.

Occurrence: 78, 142   *iii-x*   1893-2007
No status.

Halictus tumulorum (Linnaeus)

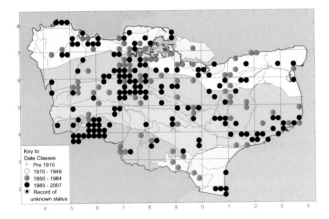

One of the most abundant bees in the county, scarcer on the clays. It frequently occurs in gardens. The phenology shows two main peaks corresponding to spring foundresses and a broader one for summer and autumn workers, males and gynes. The female is widely polylectic, as could be expected for a common bee with such a long flight period.

Occurrence: 149, 234   *iii-x*   1895-2007
No status.

Lasioglossum albipes (Fabricius)

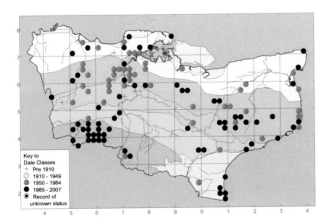

A common bee frequent on the chalk and sands. Scarcer on the clays. Occurs on the coast, including the shingle at Dungeness. The phenology shows no clear pattern. The species is widely polylectic, visiting plants such as *Ranunculus*, Rosaceae, Asteraceae, Apiaceae and *Euphorbia*.

Occurrence: 68, 121   *iii-x*   1878-2007
No status.

Lasioglossum calceatum (Scopoli)

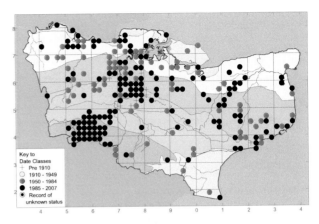

An abundant bee but scarce on clay soils. The phenology is strongly bimodal and the species is known to be eusocial. Will nest in a wide variety of situations including garden borders and chalk banks. Does not usually nest in dense aggregations in spite of its abundance. It is very widely polylectic but frequently recorded from yellow flowered Asteraceae such as *Taraxacum*.

Occurrence: 143, 230   *iii-xi*   1894-2007
No status.

Lasioglossum fratellum (Perez)

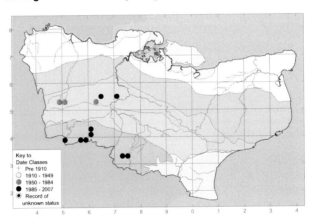

This species is found mainly on sandy heath and moors in Britain, hence is scarce and little known in the county. Only from the Lower greensand and weald sands. Thought to be primitively eusocial. Widely polylectic elsewhere but very few Kent data exist on flower visits – only known from *Taraxacum* (Asteraceae) in the county. A host of the cleptoparasitic bee, *Sphecodes hyalinatus*.

Occurrence: 9, 12   *iv-ix*   1977-2007
Kent status   Present work: pKa

Lasioglossum fulvicorne (Kirby)

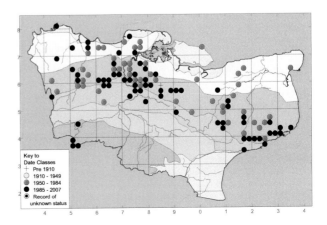

78 The Bees, Wasps and Ants of Kent

A very common bee on the chalk but with only limited data from other strata. Nesting takes place in short turf. The species is solitary, unusual for its section of the genus. The phenology is bimodal although there is only one generation per year. The first peak represents spring females and the second, summer males and females. The summer is when mating takes place and the females then hibernate until spring. The bee is widely polylectic, visiting flowers such as *Ranunculus*, Asteraceae, Rosaceae, Apiaceae and *Euphorbia*. It is the main host of the cleptoparasitic bee, *Sphecodes hyalinatus*, in the county.

Occurrence: 62, 112   iii-x   1881-2007
No status.

Lasioglossum laevigatum (Kirby)

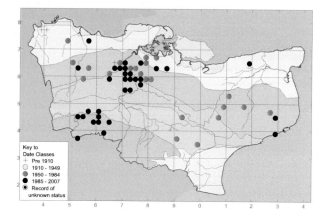

A fairly common bee but not everywhere. Particularly found on the chalk in the centre of the county but also on the sands. Only rarely found on clay. The species is solitary and nest habits are not known in the county, although it is, of course, a mining bee. Thought to be widely polylectic; flower visits are known for Rosaceae, Asteraceae, Apiaceae and Brassicaceae, many of which may represent pollen sources.

Occurrence: 32, 57   iv-x   1850s-2006
No status.

Lasioglossum lativentre (Schenck)

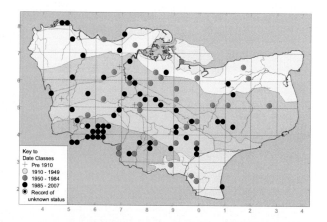

Common over much of the county although possibly a little declined. Like the preceding species, known to be a solitary mining bee. The phenology is bimodal with the second peak being particularly strong in August. Flower visits are known for Asteraceae.

Occurrence: 54, 89   ii-xi   1895-2007
No status.

Lasioglossum leucopus (Kirby)

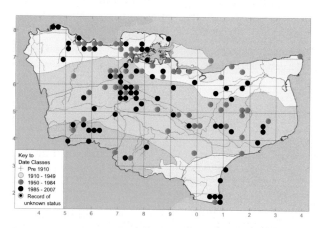

A very common bee in the county, although possibly experiencing a slight and recent decline. Particularly frequent on the chalk, sands and coastal shingle. The phenology is bimodal and the species is probably eusocial. The female nests in short turf. Flower visits are known for *Ranunculus*, Apiaceae, Asteraceae and Rosaceae.

Occurrence: 65, 124   iv-x   1897-2006
No status.

Lasioglossum leucozonium (Schrank)

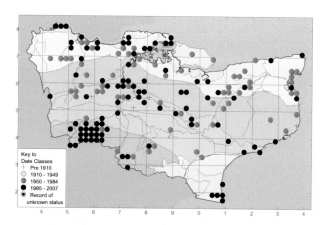

An abundant bee across much of the county, although most frequent on the sands and coastal areas. The phenology is bimodal, with strong peaks in June and August. The species is solitary. Nesting is in short turf or sparsely vegetated sandy ground. The females appear to forage mainly on Asteraceae, including both thistles and yellow flowered species.

Occurrence: 95, 168   iii-xi   1887-2007
No status.

♀ *Lasioglossum malachurum*   © Lee Manning

Lasioglossum malachurum (Kirby)

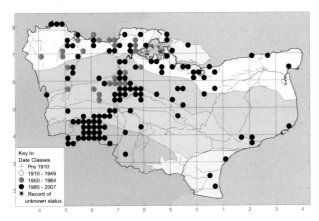

This is a eusocial mining bee that has increased enormously in range and abundance over the last three decades – apparently unknown in the county before 1966. Most frequent on the sand. With an essentially trimodal phenology, the second and third peaks being close together. Nesting takes place in hard-packed sand, often in large aggregations. The bees are frequent visitors to Asteraceae but will also visit a wide range of other plants, including *Salix* and *Ranunculus*. A host of the cleptoparasitic bee, *Sphecodes monilicornis*.

Occurrence: 117, 142   *ii-x*   1966-2007
National status   Falk: Nb
Kent status   Waite: Notable   Present work: No Kent status.

Lasioglossum minutissimum (Kirby)

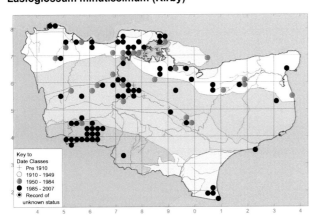

A tiny mining bee that is very common on sandy soils and coastal shingle. Much less common on clays and chalk. Nests in short turf. Probably a solitary species with a long flight period, the phenology being weakly bimodal. The female forages mainly from Asteraceae but will also visit the flowers of *Prunus spinosa*.

Occurrence: 70, 97   *iii-x*   1894-2007
No status.

Lasioglossum morio (Fabricius)

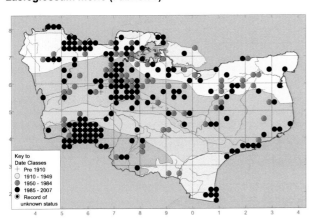

A small, metallic green, eusocial mining bee found abundantly on calcareous and sandy soils; also on the coastal shingle at Dungeness. Scarce on the weald clay. The phenology curve is essentially bimodal, with the modes quite broadly overlapping. The female usually nests in short turf or sparsely vegetated ground and the bee is frequent in gardens. Widely polylectic but often found on Asteraceae. The putative host of the cleptoparasitic bee, *Sphecodes niger*.

Occurrence: 168, 237   *iii-xi*   1887-2007
No status.

Lasioglossum nitidiusculum (Kirby)

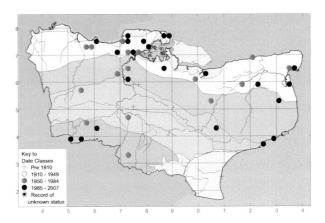

This bee seems to have declined significantly in the last two decades. Although quite widely distributed, it is most frequent in the north of the county. The records seem scattered over the strata, giving no clear indication of soil type preferences in Kent. The phenology is unclear, although it is thought to be a solitary species. The female nests in steeply sloping, south facing banks. Reported to be a host of the cleptoparasitic bees, *Sphecodes crassus*, *S. geoffrellus*, *S. miniatus* and *Nomada sheppardana*.

Occurrence: 22, 42   *iii-x*   1894-2007
No status.

Lasioglossum parvulum (Schenck)

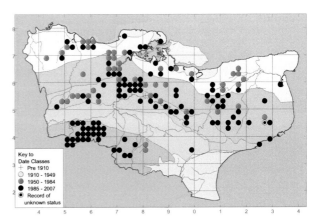

A very similar bee to the last, some females proving quite difficult to separate. An abundant bee, most often recorded from sandy soils in the west of the county but in the east, more frequent on the chalk. Like the last species, a solitary mining bee, the phenology being weakly bimodal. The nest is dug in either sloping or level ground, often in short turf. The bee is widely polylectic, although frequently visiting Asteraceae. Will also visit Rosaceae, Apiaceae and *Acer*. A putative host of the cleptoparasitic bees, *Sphecodes geoffrellus*, *Nomada sheppardana* and possibly other *Sphecodes* species.

Occurrence: 105, 156   *iii-x*   1895-2007
No status.

Lasioglossum pauperatum (Brulle)

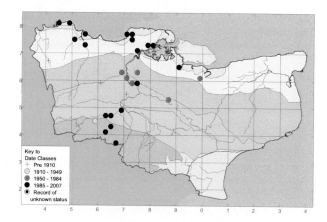

Although a rare species nationally, this mining bee is frequent enough in the county not be accorded rarity status here. Indeed, Kent seems to be a stronghold for the species. It is a small, rather nondescript, eusocial bee found over the strata to the extent that it has no clear preferences for soil type in Kent. The phenology remains unclear. The nest seems to be undescribed in the literature but is probably dug in light soils in sunny situations. Pollen sources are also unknown although flower visits are recorded for Asteraceae, Boraginaceae and Brassicaceae.

Occurrence: 19, 29   *iii-ix*   1896-2006
National status   Shirt: RDB3   Falk: pRDB3
Kent status   Waite: KRDB3   Present work: pKb

Lasioglossum punctatissimum (Schenck)

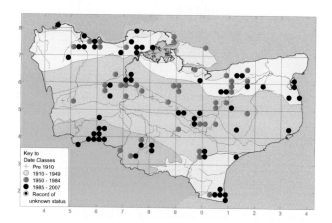

A common, solitary mining bee but possibly declining. Scarce on clay soils but otherwise fairly common. The distribution is quite strongly and positively correlated with established woodland, where the bee is often found visiting flowers in the rides. Also found on coastal sand and shingle. The phenology is unclear. Pollen is collected from Lamiaceae but the bee can also be found on yellow flowered Asteraceae, *Euphorbia* and *Epilobium*.

Occurrence: 60, 104   *iii-x*   1892-2007
No status.

Lasioglossum pauxillum (Schenck)

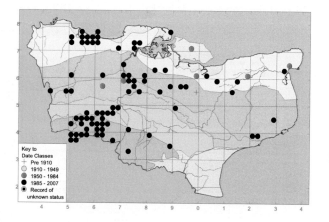

A species which has increased enormously over the last two decades and is here considered abundant. Mainly on sandy soils although occasional on the chalk and rare on the clay scarps. A small, nondescript, eusocial mining bee frequently found nesting in short turf and sparsely vegetated sandy soils in the sun. The phenology is distinctly bimodal. Probably widely polylectic, with visits recorded to such plants as: *Prunus spinosa, Ranunculus,* Asteraceae and Apiaceae.

Occurrence: 81, 88   *iv-x*   1890s-2007
National status   Falk: Na
Kent status   Waite: Notable   Present work: No Kent status.

Lasioglossum puncticolle (Morawitz)

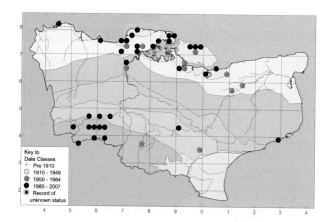

A scarce species nationally but recorded with some frequency in Kent, a stronghold. It is found locally in woodland rides, estuarine areas and coastal soft rock cliffs. It is probably solitary but the life history is not well documented. The phenology does not elucidate this. Sometimes nests in aggregations but possibly also rather solitarily. Pollen sources include Asteraceae and *Daucus carota* (Apiaceae).

Occurrence: 34, 45   *iv-x*   1969-2007
National status   Falk: Nb
Kent status   Waite: Notable   Present work: No Kent status.

Lasioglossum quadrinotatum (Kirby)

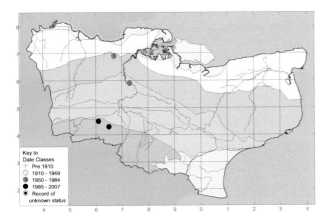

A very rare species in Kent, known widely but scarcely over England and south Wales. It is easily confused with the common *L. lativentre*. The females of the two can be very difficult to separate and the species are best identified from dissected males. *L. quadrinotatum* is probably a solitary species although no information can be given on the life history in Kent.

Occurrence: 2, 4   *iii-v*   1969-2007
National status   Falk: Na
Kent status   Present work: pKRDB1

Lasioglossum semilucens (Alfken)

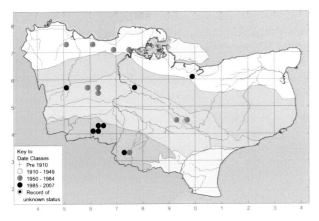

A rare species nationally, also in Kent. A tiny bee occurring only on sandy soils in the west and middle of the county. Probably a solitary species, nesting in south facing, sandy banks. Forage plants are unclear but may include yellow flowered Asteraceae and *Potentilla*.

Occurrence: 8, 18   *iv-x*   1966-2005
National status   Falk: pRDB3
Kent status   Waite: Not ranked   Present work: pKRDB3

Lasioglossum smeathmanellum (Kirby)

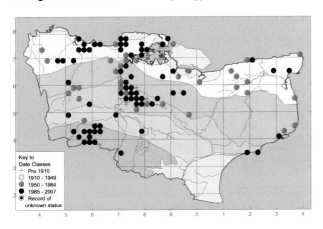

Halictinae 81

A common but small, metallic green, eusocial mining bee. The data suggest that it may be in decline, this is also being reported from other southern counties (G R Else, *pers. comm.*). The bee is most frequent on the sand but apparently absent from the coastal shingle at Dungeness. Found on chalk scarps and occasionally, coastal soft rock cliffs. The phenology is bimodal. The nest is frequently constructed in the soft mortar of old walls (probably a substitute for sandstone cliff faces) although possibly sometimes in more level ground. Flower visits are particularly for Asteraceae but also *Prunus* and Apiaceae. Males have additionally been recorded from *Chamerion angustifolium*.

Occurrence: 67, 107   *iii-x*   1893-2007
No status.

Lasioglossum villosulum (Kirby)

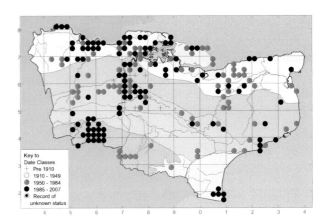

An abundant, solitary mining bee, most frequent on sandy soils, coastal shingle and the London clay, including coastal, soft rock cliffs. Also found but less frequently on chalk scarps and cliffs. The phenology is weakly bimodal. The females nest in soft soil, sometimes quite moist and clayey. Flower visits are most frequently for yellow flowered Asteraceae but also for *Smyrnium* (Apiaceae) and *Salix*.

Occurrence: 106, 197   *iii-x*   1879-2007
No status.

Lasioglossum xanthopus (Kirby)

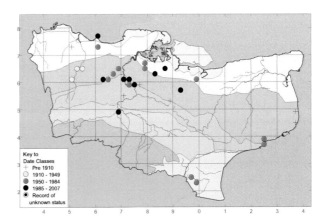

A scarce, solitary mining bee, one of the largest species of its genus in Britain. Most frequent on the chalk; rare on marsh and sandy soils. The phenology is weakly trimodal, the last mode (peaking in early October) being males, which are late flying. Flower visits are for yellow flowered Asteraceae, Brassicaceae, *Fragaria* and *Rubus* (Rosaceae), *Anthriscus* (Apiaceae) and "scabious", although pollen sources are not clear.

Occurrence: 9, 30   *iv-x*   1850s-2007
National status   Falk: pNb
Kent status   Waite: Notable   Present work: pKa

Lasioglossum zonulum (Smith)

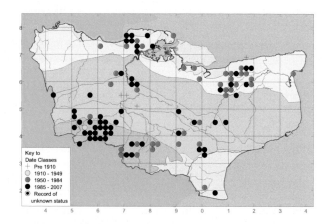

Although common, this solitary mining bee is most frequent in established woodland on sand and clay soils. Rarely coastal. The phenology is essentially bimodal with a strong second mode peaking in late August. Flower visits are known for *Ranunculus*, Asteraceae, Apiaceae and *Euphorbia*.

Occurrence: 58, 100 *iii-x* 1897-2007
No status.

Sphecodes crassus Thomson
Synonymy: *S. variegatus*

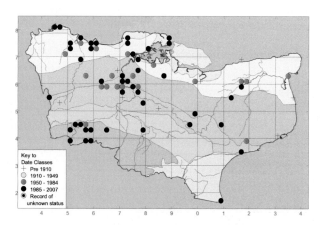

Now that good characters have been found to recognise the female of this species (the sex that is most usually found), it has been found to be common and quite widespread, requiring revision of the Kent status. A cleptoparasitic bee; the host most often cited is *Lasioglossum nitidiusculum*, but probably also *L. parvulum* is parasitised. Most frequent on sandy soils. Both sexes of *S. crassus* can be found visiting flowers for nectar but the females have no scopal hairs for transporting pollen. Recorded flower visits are for Asteraceae and Apiaceae, the latter most often for males.

Occurrence: 44, 71 *iv-ix* 1896-2006
National status Falk: pNb
Kent status Waite: Notable Present work: No Kent status.

Sphecodes ephippius (Linnaeus)
Synonymy: *S. divisus, S. similis*

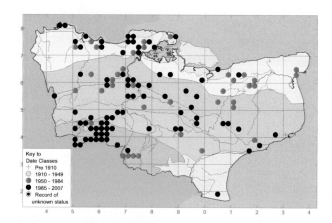

An abundant cleptoparasite with several hosts, including *Lasioglossum leucozonium, L. zonulum* and *L. lativentre*; smaller examples may have come from nests of *L. nitidiusculum* or *L. parvulum*. Found on most soil types including the London clay of the Blean. Flower visits are for Asteraceae, Apiaceae and *Euphorbia*.

Occurrence: 81, 124 *iii-x* 1889-2007
No status.

Sphecodes ferruginatus von Hagens

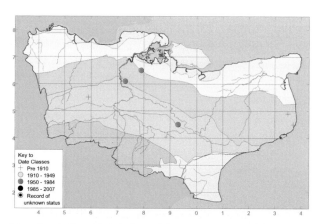

A rare bee in the county but possibly over looked at times. A recent report of *S. ferruginatus* from Swanscombe (Allen, 2006) proved incorrect. Hosts may possibly include *Lasioglossum fulvicorne*, a common mining bee on the chalk. I have no information on flower visits in the county.

Occurrence: 0, 6 *vi-viii* 1897-1979
National status Falk: pNb
Kent status Waite: Notable Present work: pKRDB1

Sphecodes geoffrellus (Kirby)
Synonymy: *S. fasciatus, S. affinis*.

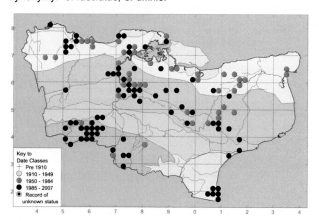

A very common cleptoparasite. The cited host is *Lasioglossum nitidiusculum* but the dot map suggests that *L. parvulum* may now be the usual host in the county. Other hosts are suspected in certain instances. Most frequent on sand and chalk. Flower visits are recorded for Asteraceae and Apiaceae.

Occurrence: 76, 119   *iii-x*   1893-2007
No status.

Sphecodes gibbus (Linnaeus)

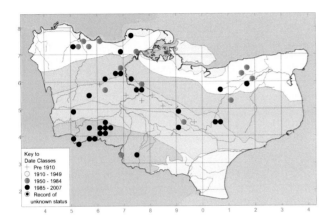

A fairly common bee, cleptoparasitic on *Halictus rubicundus* and probably some *Lasioglossum* species. Most frequent on sandy soils and in river valleys. Will visit Asteraceae such as mayweeds and thistles, and possibly *Reseda*.

Occurrence: 29, 49   *iv-x*   1895-2005
No status.

Sphecodes hyalinatus von Hagens

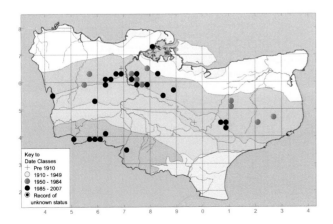

Frequently found on calcareous grassland and occasionally on the Weald sands, rare elsewhere. A cleptoparasite of principally *Lasioglossum fulvicorne* in the county and occasionally found on *L. fratellum*. The male visits Asteraceae and Apiaceae; a female was recorded from *Euphorbia cyparissias*.

Occurrence: 24, 36   *iv-x*   1899-2006
No status.

Sphecodes longulus von Hagens

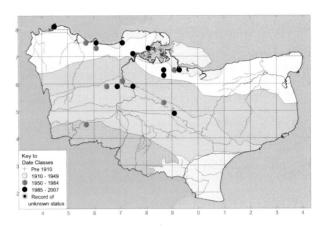

Particularly recorded in Kent by Dr G H L Dicker in the 1980s, only infrequently found before and since. A cleptoparasite of *Lasioglossum minutissimum* and possibly other *Lasioglossum* species. The habitat is sandy soils and sometimes soft rock cliffs. Some strata may be particularly favoured. Flower visits are recorded for Asteraeae and Apiaceae.

Occurrence: 11, 19   *iv-x*   1904-2003
National status   Falk: pNa
Kent status   Waite: Notable   Present work: pKa

Sphecodes miniatus von Hagens
Synonymy: *S. dimidiatus*

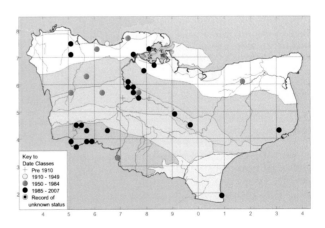

Quite frequently recorded in the county but possibly sometimes confused with the similar, very common *S. geoffrellus*. There is a westerly bias to the data and the species is most often recorded on the sands. The recorded hosts are *Lasioglossum nitidiusculum* and *L. parvulum*, but a female *S. miniatus* was taken coursing along and inspecting holes in an old stone wall in which *L. smeathmanellum* was nesting (*pers. obs.*). This is a possible additional host. Known to visit Asteraceae.

Occurrence: 23, 35   *iv-ix*   1894-2006
National status   Falk: pNb
Kent status   Waite: Notable   Present work: No Kent status.

Sphecodes monilicornis (Kirby)
Synonymy: *S. subquadratus*

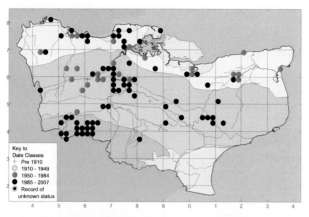

A common cleptoparasitic bee, hosted by *Halictus rubicundus*, *Lasioglossum malachurum* and possibly also other halictines. Mainly distributed on the sands; uncommon on other soil types. Visits mainly Asteraceae; males occasionally also found on Apiaceae.

Occurrence: 68, 95   iv-x   1895-2007
No status.

Sphecodes niger von Hagens

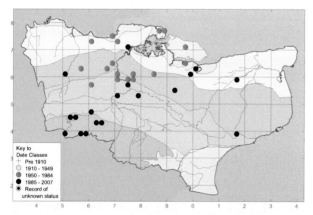

Although a nationally rare species, *S. niger* is quite frequently recorded in Kent, a stronghold for the species. Originally recorded mostly on the chalk but there are now many data for sandy soils. The putative host is *Lasioglossum morio*, an abundant bee in the county. The cuckoo is often recorded on Asteraceae.

Occurrence: 18, 35   iv-x   1976-2006
National status   Shirt: RDB3   Falk: pRDB3
Kent status   Waite: Not ranked   Present work: pKb

Sphecodes pellucidus Smith

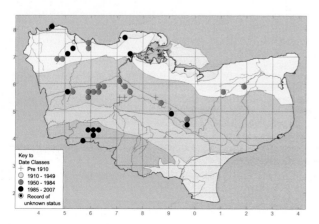

This species is given scarcity status here, there being a possible decline in the county. Most frequent in May and June, with a second mode in late August. Like its host, *Andrena barbilabris*, this species is almost entirely confined to sandy soils. The few flower visit data available suggest that Asteraceae is the main host plant family.

Occurrence: 13, 36   iv-ix   1892-2005
Kent status   Present work: pKb

Sphecodes puncticeps Thomson

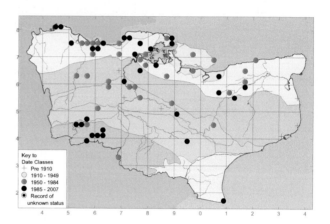

A fairly common cleptoparasitic bee most often recorded on sandy soils. The species is most frequently recorded in August and is mainly hosted by *Lasioglossum lativentre*, a common mining bee. Flower visits are recorded for Asteraceae, including mayweeds, thistles, *Solidago* and *Pulicaria dysenterica*.

Occurrence: 31, 69   v-x   1897-2004
No status.

Sphecodes reticulatus Thomson

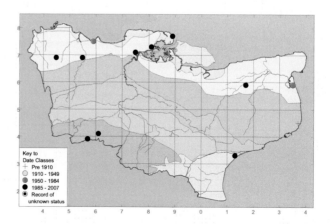

A very scarce bee in the county quite often found on the coast. There are no firm observations on host/parasite relationships but it is likely that the second brood of *Andrena dorsata* forms the main host, with *A. barbilabris* a possible alternative. Males have been recorded as visiting Asteraceae and Apiaceae.

Occurrence: 9, 12   v-ix   1890s-2006
National status   Shirt: RDB3   Falk: pNa
Kent status   Waite: Notable   Present work: pKa

Halictinae to Melittinae

Sphecodes rubicundus von Hagens
Synonymy: *S. rufiventris* (misident.), *S. ruficrus* (misident.)

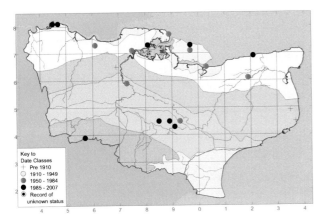

Scarce in the county and possibly declining. Found on coastal soft rock cliffs and also inland on the sands, chalk and Weald clay. The host is likely to be *Andrena labialis*, a common mining bee. Flower visits are for Asteraceae.

Occurrence: 9, 23   *v-vi*   1895-2006
National status   Falk: pNa
Kent status   Waite: Notable   Present work: pKa

Sphecodes scabricollis Wesmael

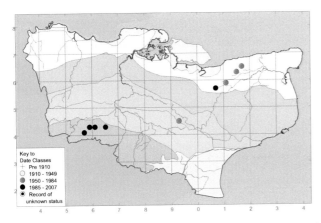

A rare cleptoparasitic bee distributed almost entirely in established woodland, including on the London clay of the Blean, and High Weald sands. The host is believed to be the mining bee, *Lasioglossum zonulum*, frequent in such habitats. Males have been taken from *Pulicaria dysenterica* and *Cirsium arvense*.

Occurrence: 5, 9   *vi-ix*   1970-2005
National status   Shirt: RDB3   Falk: pRDB3
Kent status   Waite: KRDB2   Present work: pKRDB3

Sphecodes spinulosus von Hagens

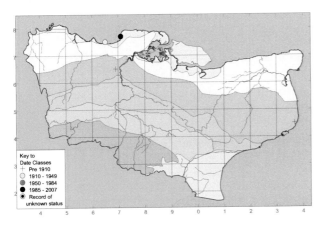

A very rare cleptoparasitic bee; the host is believed to be *Lasioglossum xanthopus*. The records for Ham Street Woods NNR from 1990 have been discarded as unreliable. Most data are coastal or estuarine, although old records exist for Upper Halling. British flower visits are recorded for *Euphorbia* and Apiaceae.

Occurrence: 1, 4   *vi-viii*   1898-2006
National status   Shirt: RDB3   Falk: pRDB2
Kent status   Waite: KRDB1   Present work: pKRDB1

Subfamily Melittinae

♀ *Melitta leporina*

This is the last subfamily of short tongued bees; it is possible that the long tongued bee subfamilies were evolutionarily derived from the same immediate ancestral group as the melittines. Five of the six British Melittinae occur in Kent. The species tend to be local or scarce and are all mining bees.

The melittines have a strong tendency to oligolecty, i.e. they forage on a narrow range of related plants. Thus *Melitta leporina* forages mainly on *Trifolium*, *M. haemorrhoidalis* on *Campanula* and *M. tricincta* on *Odontites vernus*.

The British melittines can be placed in two or three tribes: the Melittini might not include *Macropis*, depending on choice of classification. *Dasypoda* is more distinct and placed in the Dasypodaini.

Tribe Melittini.

♀ *Macropis europaea*

I have included *Macropis* in this tribe, which thus contains all the British species apart from *Dasypoda hirtipes*.

The Bees, Wasps and Ants of Kent

Melitta haemorrhoidalis (Fabricius)

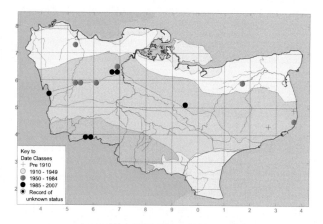

A rare mining bee in the county, most often recorded on the chalk scarps but occasionally on sandy soils. The female is oligolectic on *Campanula* species, being recorded from *C. rotundifolia*, *C. trachelium* and others.

Occurrence: 6, 15 *vi-viii* 1850s-2005
Kent status Present work: pKRDB3

Melitta leporina (Panzer)

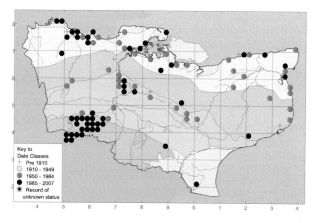

A common bee with a fairly short flight period. The female predominantly visits *Trifolium repens* for pollen but will also exploit related plants such as *Melilota* spp., *Medicago sativa* and possibly other *Medicago* species. Frequent on sandy soils. The main host of the cleptoparasitic bee, *Nomada flavopicta*.

Occurrence: 50, 95 *vi-ix* 1850s-2007
No status.

Melitta tricincta Kirby

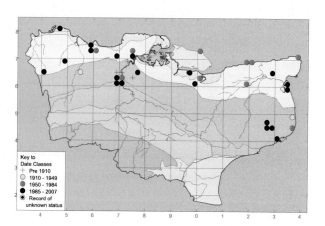

A nationally scarce species but not uncommon in Kent, hence not classified. It is closely related to the preceding species but the female visits solely *Odontites vernus* for pollen. It also has a later but over lapping flight period. Most often recorded on the chalk. A possible host of the cleptoparasitic bee, *Nomada flavopicta*.

Occurrence: 20, 37 *vii-ix* 1890s-2007
National status Falk: pNb
Kent status Waite: Notable Present work: No Kent status.

Macropis europaea Warncke

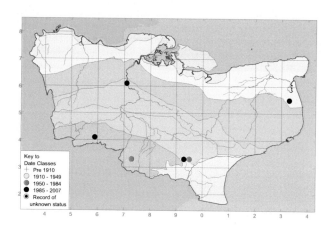

A rare mining bee that visits only *Lysimachia vulgaris* for pollen. *Lysimachia*, however, does not produce nectar but floral oils. The female provisions these with the pollen. The adults will visit *Rubus* and other plants for nectar before the *Lysimachia* is in flower. Soil preferences are unclear.

Occurrence: 4, 6 *vi-ix* 1978-2005
National status Shirt: RDB3 Falk: pNa
Kent status Waite: Notable Present work: pKRDB3

Tribe Dasypodaini.

♀ *Dasypoda hirtipes*

This tribe contains one British species, *Dasypoda hirtipes*, a rather local bee in Kent.

Melittinae to Megachilinae

Dasypoda hirtipes (Fabricius)
Synonymy: *D. altercator, D. plumipes*

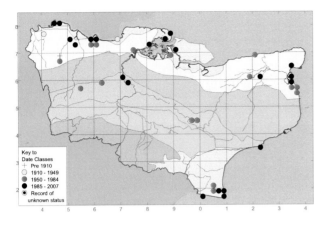

A very distinctive mining bee, widespread on sandy soils but apparently absent from the High Weald sands in the county. The nests are dug in light sandy soils, often in small aggregations. An oligolege of Asteraceae, frequently the yellow flowered species.

Occurrence: 21, 40 *vi-ix* 1850s-2007
National status Falk: pNb
Kent status Waite: Notable Present work: No Kent status.

Subfamily Megachilinae

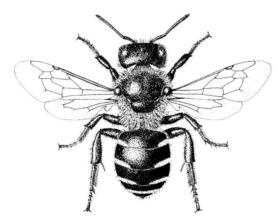

♀ *Hoplitis claviventris*

A large, diverse subfamily, the first of the long tongued bees. All British species have two submarginal cells in the fore wing, a rectangular labrum longer than broad and, in the independent species, the female has the pollen carrying scopa solely on the underside of the gaster. The British species are all assigned to the tribe Megachilini, which has three subtribes: Anthidiina, Osmiina and Megachilina. There are two British cleptoparasitic genera, *Stelis* (Anthidiina) and *Coelioxys* (Megachilina). These are scarce compared with *Sphecodes* and *Nomada*, where some of the species may be abundant. Many free living species of Megachilinae are aerial nesters, some exceptions being *Megachile dorsalis*, *M. maritima*, and sometimes *M. willughbiella*. A few *Osmia* nest in empty snail shells. This is the only subfamily of solitary bees where the female gathers nesting materials, including mud, cut leaves, leaf mastic, resin or plant fibres. Some species forage on Asteraceae and Rosaceae, whilst several *Osmia* are found on *Lotus* and other Fabaceae.

Tribe Megachilini. A discussion of the three subtribes follows:

Subtribe Anthidiina: This subtribe has two British genera, *Anthidium* and *Stelis*, with a total of five UK and Kent species. *Anthidium* constructs nests of plant fibres within suitable cavities; the hairs from the underside of lambs ear are frequently used. *Stelis* are cleptoparasites of Osmiina and *Anthidium manicatum* in the British fauna but abroad other groups may be used by non-British species.

Subtribe Osmiina: The classification of the osmiines is in a state of flux and the generic assignment of *Osmia spinulosa* is provisional. Given this placement, there are four Kent genera. Osmiines build nests with a variety of materials, leaf mastic being frequently used. *Osmia rufa* uses mud, unusual in the British fauna but a practice that gave rise to the vernacular name "mason bees" for the genus. The nests of most species are built in a variety of tubular cavities such as hollow stems and old snail shells, but some, mud dauber-like, will construct mud nests to fit a larger cavity. *Stelis* and Sapygidae are frequent cleptoparasites of osmiines. *Chrysura radians* is a parasitoid of *Osmia leaiana*.

Subtribe Megachilina: There are two British and Kent genera in this subtribe, *Megachile* and *Coelioxys*. All British *Megachile* are leaf cutters, using the cut leaves to build the cells of their nests. Roses, including cultivated varieties, are a frequent source for leaves. *Megachile* species elsewhere may use resin or mud for nest construction, although the generic assignment of these bees is still in question. A species in this group, one that uses mud rather than resin, *M. parietina*, has been captured once in the county. However, it is not thought to be native. *Coelioxys* are cleptoparasites, laying their eggs in the nests of *Megachile* but the genus *Anthophora* is also parasitised.

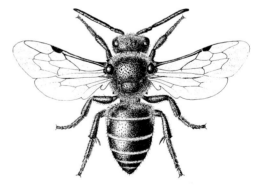

♀ *Stelis punctulatissima*

Anthidium manicatum (Linnaeus)

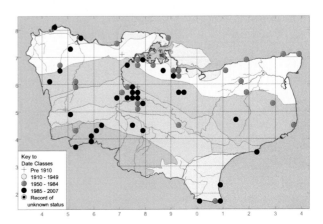

A fairly common bee, distinctive in having paired yellow spots on the sides of the gaster. Most frequent on the sands but also found on the chalk and clays. Quite often coastal, and known from Dungeness. The male is territorial, patrolling and disabling or killing other bees flying near the flowers in its territory. The female builds nests in tubular cavities from plant fibres, often gathered from the underside of the leaves of *Stachys byzantina* and has hence gained the name of "wool carder bee". Flower visits are made to Lamiaceae, Fabaceae and probably Boraginaceae.

Occurrence: 32, 61   *vi-ix*   1879-2007
No status.

Stelis breviuscula (Nylander)

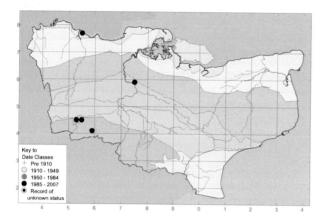

A rare cleptoparasite but spreading with its host species, *Heriades truncorum*. First recorded from the UK in 1984 from West Sussex (coll. G R Else) and now known sparingly in the south-east. Its distribution will mirror *Heriades* which at present is found mainly on the sands. Flower visits are known for yellow flowered Asteraceae, particularly *Senecio*.

Occurrence: 5, 5   *vii-viii*   2000-2007
National status   Shirt: RDB1   Falk: pRDBK
Kent status   Present work: pKRDB3

Stelis ornatula (Klug)

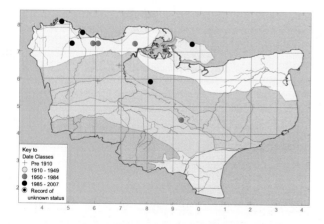

Formerly more common and possibly declining. Not known south of the lower greensand in the county and present on the sands, chalk and soft rock cliffs. The host in the UK is *Hoplitis claviventris*. It has been recorded from the flowers of *Potentilla* and Asteraceae, and abroad, also: *Lotus*, *Veronica* and *Rubus*.

Occurrence: 5, 11   *v-viii*   1899-2001
National status   Shirt: RDB3   Falk: pRDB3
Kent status   Waite: KRDB2   Present work: pKRDB3

Stelis phaeoptera (Kirby)

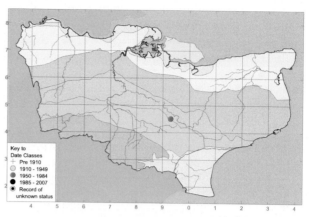

A very rare bee and probably declining. Recorded in Kent only from the Lower greensand. The recorded hosts in the UK are *Osmia caerulescens* and *O. leaiana*, two common solitary bees. On the continent, several other osmiines and *Anthidium manicatum* are recorded. British flower visits are known for *Lotus*, *Veronica* and Asteraceae.

Occurrence: 0, 2   *vi*   1879, 1970
National status   Shirt: RDB3   Falk: pRDB2
Kent status   Waite: KRDB1   Present work: pKRDB1

Stelis punctulatissima (Kirby)

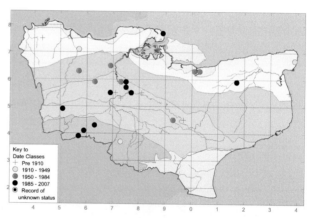

The most common species of its genus in the UK and Kent but still a scarcity. Widely distributed on the sands and chalk, rarer on clays. The hosts are *Anthidium manicatum* and probably *Osmia leaiana*. Flower visits are recorded for *Malva*, *Lotus*, *Veronica*, *Rubus* and Asteraceae.

Occurrence: 10, 26   *vi-viii*   1850s-2007
National status   Falk: pNb
Kent status   Waite: Notable   Present work: pKa

Heriades truncorum (Linnaeus)

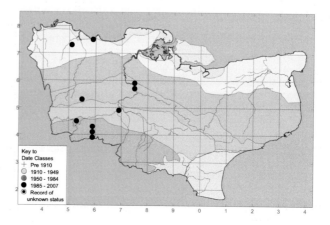

Megachilinae

A scarce species but spreading. A recent addition to the Kent fauna and so far recorded mostly on sandy soils. Nests in existing burrows in dead wood; the cells and entrance are sealed with resin including, but not exclusively, that of conifers such as *Pinus sylvestris*. Flower visits are known mainly for Asteraceae; *Senecio jacobaea* may represent the main pollen source. The obligate host for *Stelis breviuscula*.

Occurrence: 10, 10 *vi-ix* 1997-2007
National status Shirt: RDB3 Falk: pRDBK
Kent status Waite: KRDB1 Present work: pKa

Chelostoma campanularum (Kirby)

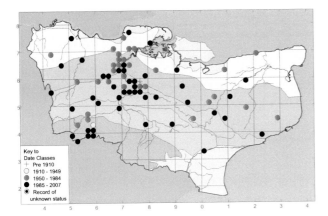

A small but common bee frequently nesting in beetle exit holes in dead wood. Although of similar habitus to the last species, this genus is closer to *Osmia* and *Hoplitis*. Recorded particularly from the chalk, Lower greensand and High Weald sands, but present on other soil types. As the species name suggests, there is a relationship with *Campanula*, which provides a large percentage of pollen collected. However, flower visits have also been recorded for Asteraceae and other plant families. These may possibly represent only nectar sources.

Occurrence: 44, 76 *vi-ix* 1897-2007
No status.

Chelostoma florisomne (Linnaeus)

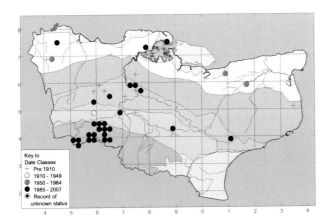

A larger bee than the previous species but rather less common. Most frequently recorded on the weald sands but rare elsewhere; usually on sands. The nest is formed in beetle burrows in dead wood. The main forage plants appear to be *Ranunculus* species but the female has also been recorded from *Veronica* and *Rosa* in the county. A frequent host of *Sapyga quinquepunctata* and *Monosapyga clavicornis*.

Occurrence: 27, 41 *v-vii* 1893-2007
No status.

Osmia aurulenta (Fabricius)

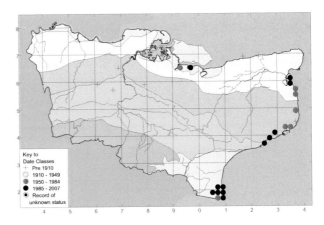

A rather uncommon bee in the county and modern records are strictly from coastal areas, including the shingle at Dungeness. The nest is constructed in empty snail shells. There are few data on forage plants in the county but include *Armeria maritima*. A possible host of *Sapyga quinquepunctata*.

Occurrence: 12, 26 *iv-viii* 1895-2006
Kent status Present work: pKb

Osmia bicolor (Schrank)

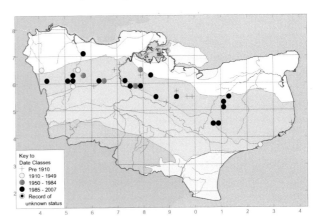

Possibly declining and found only on calcareous grassland in the county. Another nester in empty snail shells. Flower visits have been recorded for *Salix*, *Lotus*, *Viola*, Apiaceae, *Euphorbia amygdaloides* and *Ajuga reptans*.

Occurrence: 17, 34 *iii-vii* 1893-2007
National status Falk: pNb
Kent status Waite: Notable Present work: pKb

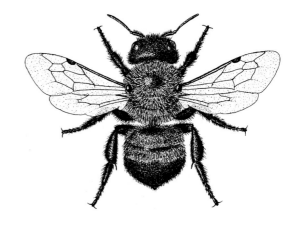

♀ *Osmia pilicornis*

Osmia caerulescens (Linnaeus)

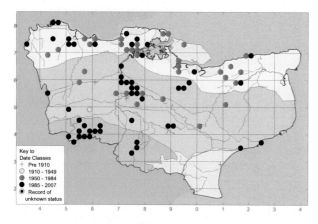

A common bee closely related to *O. leaiana* and *O. niveata*, from which the males are not always easily distinguished. Nests in beetle exit holes in dead wood and similar situations, the cells and entrance being sealed with leaf mastic prepared with salivary secretions. Forages on *Lotus, Salvia, Nepeta, Veronica, Rubus fruticosus* agg. etc. A possible host of *Sapyga quinquepunctata*.

Occurrence: 51, 89   *iv-ix*   1878-2007
No status.

Osmia leaiana (Kirby)

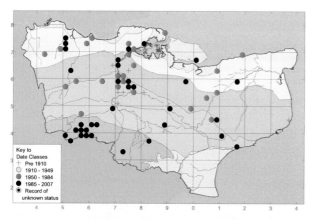

Common but not as frequent as the last species. Very similar in habits to the last. Nests in dead wood and forages on Asteraceae and *Ballota*. A frequent host of *Chrysura radians, Sapyga quinquepunctata* and *Monosapyga clavicornis*.

Occurrence: 34, 65   *iv-viii*   1892-2007
No status.

Osmia niveata (Fabricius)
Synonymy: *O. fulviventris*

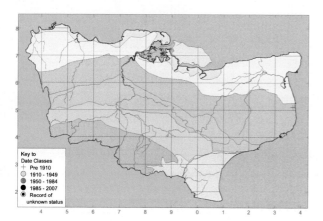

A specimen dating from the mid-nineteenth century from near Walmer was recently determined (G R Else) as this species. It is possible that it was a vagrant. The habits of this bee are probably very similar to the previous two.

Occurrence: 0, 1   *A Kent flight period cannot be defined*   1850
Kent status   Present work: Probably not native, therefore no status.

Osmia pilicornis Smith

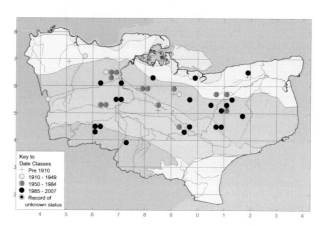

This bee is given scarcity status, although somewhat borderline. It is a species of broad-leaved woodland rides and glades, where it particularly visits *Ajuga reptans* and sometimes *Nepeta* and *Vicia*. Nests are thought to be in existing holes in dead wood.

Occurrence: 21, 42   *iii-vii*   1850s-2005
National status   Shirt: RDB3   Falk: pNa
Kent status   Waite: Notable   Present work: pKb.

Osmia rufa (Linnaeus)

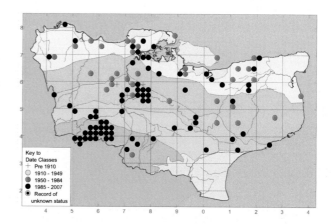

A very common bee found widely throughout the county, except in some low lying marshy areas. It is most frequent on sandy soils and scarcest on the weald clay. It will nest in any suitable cavity, including holes in masonry, hollow plant stems, beetle exit holes in dead wood and easily adapts to trap nests. The cells and entrance are sealed with mud mixed with salivary secretions. The female forages on a wide range of plants, including rosaceous fruit trees, Lamiaceae, Boraginaceae and Asteraceae.

Occurrence: 75, 112   *iii-viii*   1892-2007
No status.

Megachilinae

Osmia spinulosa (Kirby)
Synonymy: *Hoplitis (Anthocopa) spinulosa, Hoplosmia spinulosa*

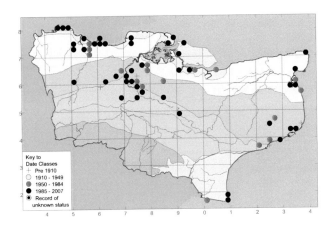

Although common, this bee is rather restricted in range. Absent from south of the Lower greensand, except found on the shingle at Dungeness. The female nests in empty snail shells and forages on Asteraceae.

Occurrence: 43, 64   *v-ix*   1895-2007
No status.

Hoplitis claviventris (Thomson)

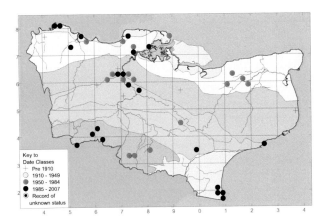

A widely distributed bee without discernible preference for soil type. The bee frequently nests in bramble and rose stems and can be reared from such. There are few Kent data for flower visits but the female does visit *Lotus*. Host of *Stelis ornatula*.

Occurrence: 21, 44   *v-ix*   1850s-2007
No status.

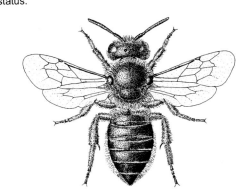

♀ *Megachile centuncularis*

Osmia xanthomelana (Kirby)

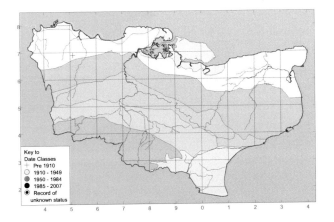

Presumably long extinct in the county and only known from one very old record, from Birch Wood (coll. F Smith). The modern British distribution is very much on coastal soft rock cliffs and in only a few localities. The female constructs nests with occasionally up to six vertical pot shaped cells, of mud mixed with small pebbles. The lower part of each cell is constructed in the soil, occasionally the whole cell. Pollen sources particularly include *Lotus corniculatus* and *Hippocrepis comosa*.

Occurrence: 0, 1   *No Kent flight period can be given*   1850s
National status   Shirt: RDB1   Falk: pRDB1
Kent status   Present work: pKRDB1+

Megachile centuncularis (Linnaeus)

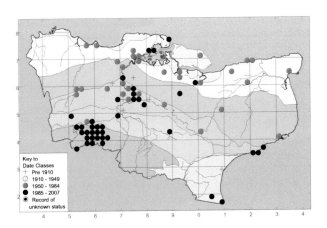

A common leaf cutter bee with records scattered over much of the county. Most data are from sandy soils but this could in part be due to observer bias. The bee is often found in gardens. The nest is constructed in beetle exit holes in dead wood and in hollow pithy stems, and the bee can adapt to some sorts of trap nests. The cells are constructed from cut leaves, frequently those of roses, including cultivated ones. There are few Kent data on flower visits but *Rubus fruticosus* agg. and Asteraceae are used. Host of more than one *Coelioxys*.

Occurrence: 40, 78   *iv-ix*   1893-2007
No status.

Megachile circumcincta (Kirby)

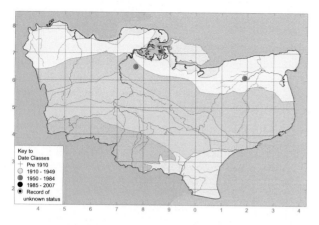

A very rare bee in Kent with no modern data. The species may also be in decline in other southern counties, although no reason can be given. It nests chiefly in the soil of sandy dunes and often visits *Lotus corniculatus* for pollen.

Occurrence: 0, 3   *vi-vii*   1895-1978
Kent status   Present work: pKRDB1

Megachile dorsalis Perez
Synonymy: *M. leachella, M. argentata*

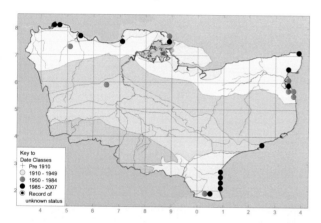

A scarce bee in the county with modern records entirely confined to coastal sand and shingle. There are older records inland from sandy soils. Nesting occurs in dry, sandy soils and the cells are constructed from cut leaves. The female has been recorded foraging from *Rubus*, *Sedum* and *Armeria maritima*.

Occurrence: 14, 22   *vi-viii*   1850s-2007
National status   Falk: pNb
Kent status   Waite: Notable   Present work: pKb

Megachile ligniseca (Kirby)

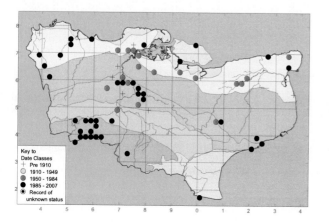

A fairly common leaf cutter but mostly on the sands. The female is large and powerful enough to bore into hard, dead wood where she constructs her cells of cut leaves. Pollen is collected mostly from Asteraceae, including thistles, *Centaurea,* and *Arctium* but also *Lathyrus* (Fabaceae).

Occurrence: 39, 65   *vi-ix*   1896-2007
No status.

Megachile maritima (Kirby)

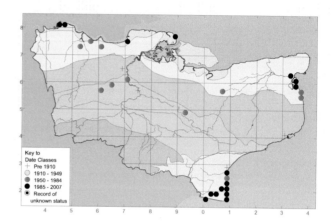

A scarce bee in the county with modern records entirely confined to coastal sand and shingle. Formerly found inland. The nest is constructed of cut leaves in dry, sandy soil and the female visits mainly Asteraceae and *Rubus*. An unusual inland record is for *Ballota*. The host for the cleptoparasitic bee, *Coelioxys conoidea*.

Occurrence: 16, 31   *vi-viii*   1896-2006
Kent status   Present work: pKb

Megachile versicolor Smith

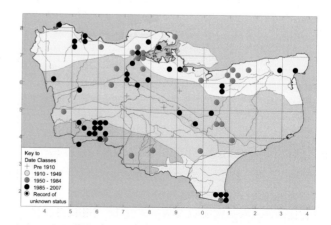

A leaf cutter almost as frequent as *M. centuncularis*. There is a loose but positive correlation between the distribution of this bee and established woodland although it also occurs in coastal sites. The nest is constructed in dead wood and pithy stems and the species adapts well to trap nests. The female forages on Asteraceae, *Rubus* and Fabaceae.

Occurrence: 40, 71   *v-ix*   1897-2007
No status.

Megachilinae 93

Megachile willughbiella (Kirby)

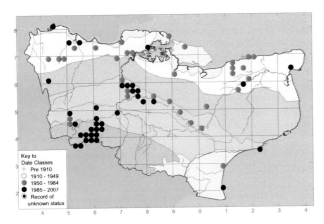

A fairly common bee found on sandy soils and less frequently on the clays. It occurs in gardens where the female may nest in plant pots. Although the nest is usually constructed in the soil the female possibly also builds in beetle exit holes in dead wood. Rose leaves are frequently used for cell construction. The female forages on Asteraceae such as *Centaurea* and also on other plants e.g. *Campanula*. Host of the cleptoparasitic bees, *Coelioxys quadridentata*, *C. elongata* and *C. rufescens*.

Occurrence: 36, 71   *v-viii*   1878-2007
No status.

Coelioxys conoidea (Illiger)
Synonymy: *C. vectis*

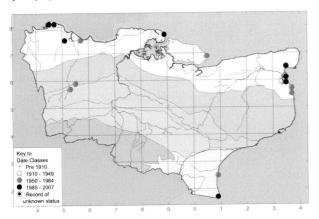

As with all *Coelioxys*, not a common bee in the county. The distribution is mainly coastal, as is that of the usual host, *Megachile maritima*. The limited inland data are from the Lower greensand. Flower visits are known for Asteraceae and *Convolvulus*.

Occurrence: 8, 15   *vi-viii*   1952-2006
Kent status   Present work: pKa

Coelioxys elongata Lepeletier

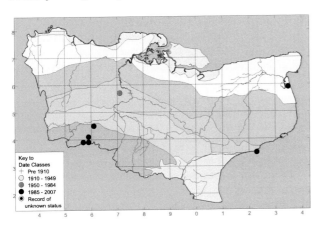

Very scarce but possibly under recorded and a bee similar in appearance to *C. inermis*. Most modern data are for the Tunbridge Wells area, apart from two coastal records. The recorded host is *Megachile willughbiella* but probably also used is *M. centuncularis*.

Occurrence: 6, 8   *vi-vii*   1906-2006
Kent status   Present work: pKa

Coelioxys inermis (Kirby)

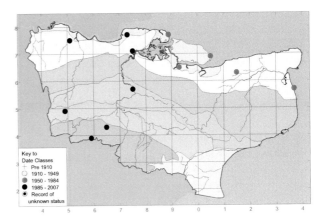

Slightly more frequent than the last species but still scarce. Records scattered over much of the county but apparently not for the southeast. The data are for various strata including coastal sand. A flower visit is for *Senecio jacobeae* (Asteraceae). The recorded host is *Megachile centuncularis*.

Occurrence: 8, 12   *vi-ix*   1878-2003
Kent status   Present work: pKa

Coelioxys mandibularis Nylander

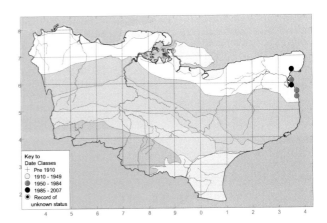

A rare cleptoparasitic bee confined in Kent to a small stretch of coastline between Deal and Cliffsend. A record for Grain was a misidentification for *C. inermis* (C Clee, *pers. comm.*). The main recorded habitat is coastal sand dunes and loose sandy areas. The Kent form, which is smaller than that from Merseyside and Wales, is probably hosted by *Megachile dorsalis*. Flower visits are not known for the UK but probably include *Lotus* and Asteraceae.

Occurrence: 2, 5   *vi-viii*   1973-2006
National status   Shirt: RDB3   Falk: pRDB3
Kent status   Waite: KRDB2   Present work: pKRDB2

Coelioxys quadridentata (Linnaeus)

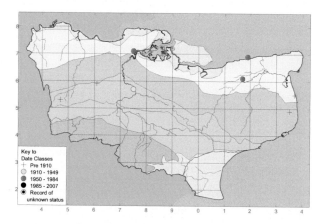

A rare bee showing a distinct decline in recent years. The data, mostly old, are scattered over much of the county but not south of the Lower greensand. The recorded host is *Megachile willughbiella*.

Occurrence: 0, 8   *vi-vii*   1850s-1979
National status   Shirt: RDB3   Falk: pRDB3
Kent status   Present work: pKRDB1

Coelioxys rufescens Lepeletier & Serville

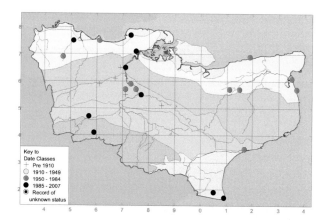

One of our more common *Coelioxys* but apparently showing a decline. Mostly recorded from sandy soils, including alluvial sands. The hosts are *Megachile* spp. but apparently also *Anthophora bimaculata*. The female has been found inspecting holes in posts containing *M. centuncularis* nests. Flower visits are known for *Senecio* (Asteraceae).

Occurrence: 9, 25   *vi-ix*   1896-2006
Kent status   Present work: pKa

Subfamily Apinae

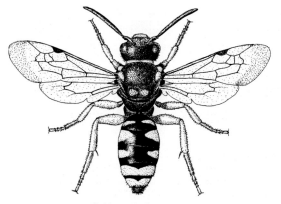

♀ *Nomada flavopicta*

As treated here, this subfamily also includes those tribes assigned to the former subfamilies Anthophorinae and Xylocopinae. Whilst Nomadini and Xylocopini are recognised here as tribes separate from the Apini, the remainder of "anthophorines" are better treated as subtribes within the last named tribe. Whilst this classification may seem controversial, it does show the true relationships within the subfamily. The number of ovarioles in the female ovaries is constantly 4 in the subfamily Apinae (except this is augmented in some of the corbiculate bees), whilst in all other subfamilies of bees the number is 3. This is a convincing synapomorphy for the Apinae as defined here. The British subtribes within Apini are: Eucerina, Anthophorina, Bombina and Apina.

Nomadini (*Nomada* and *Epeolus*) are exclusively cleptoparasites whilst, on a world wide basis, Xylocopini and Apini both comprise a range of solitary, social and cuckoo bees.

Tribe Xylocopini.

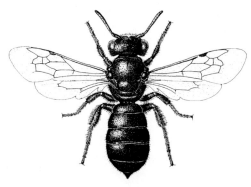

♀ *Ceratina cyanea*

This tribe has one British species, *Ceratina cyanea*, and a further, vagrant species, *Xylocopa violacea*, that appears to be establishing. The two genera are subtribally separated.

Xylocopa violacea (Linnaeus)

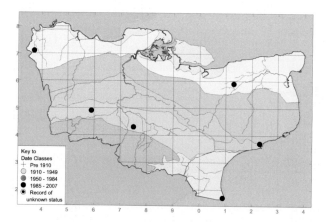

This is a very large bee superficially resembling a large *Bombus* queen but all black with a strong violaceous colouration to the wings. It has congeners that are communal. The bee has recently been recorded as successfully nesting in the UK and at least one of the Kent records is believed to be of over-wintering bees. *X. violacea* is a "large carpenter bee" and can chew into sound, dead wood to make its nest. The female will visit a variety of flowers.

Occurrence: 6, 6   *iv-viii*   1999-2007
Kent status   Present work: pKRDBK

Ceratina cyanea (Kirby)

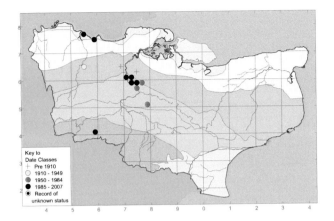

Called the "blue carpenter bee" this is a much smaller, native species that frequently nests in hollow pithy stems and from which adults have been recovered. It is a rare and local species recorded mainly from the chalk and lower greensand. It is interesting because adults can be found at all times of the year – in winter within bramble stems etc. Further, several adults may be found in one stem, possibly indicating communal behaviour. The female visits Asteraceae and *Rubus*, amongst other flowers.

Occurrence: 7, 14   *i-xii*   1861-2006
National status   Shirt: RDB3   Falk: pRDB3
Kent status   Waite: KRDB3   Present work: pKRDB3

Nomada argentata Herrich-Schaeffer

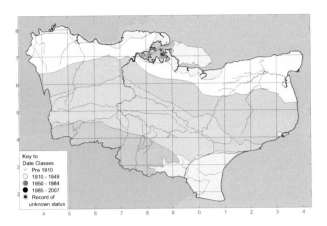

A very rare bee in the county and cleptoparasitic on the mining bee *Andrena marginata*, itself rare in Kent. If this *Nomada* is still present in Kent, it will only be found on chalk grassland where scabious species grow, food plants of *A. marginata*. *N. argentata* also visits scabious.

Occurrence: 0, 1   *viii*   1918
National status   Falk: pRDB3
Kent status   Present work: pKRDB1+

Tribe Nomadini.

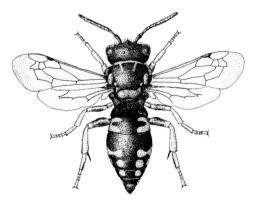

♀ *Epeolus variegatus*

This tribe contains two British genera, of which some *Nomada* can prove common, even abundant, bees. The hosts of most *Nomada* are mining bees of the genus *Andrena*, but a few species are found on *Lasioglossum*, *Melitta* and *Eucera*. The single British *Nomada* on this last genus, *N. sexfasciata*, is extremely rare and not known from Kent. 23 *Nomada* species have been recorded from the county. The genus *Epeolus*, with two British and Kent species, proves scarcer and is hosted by *Colletes* mining bees. *Nomada* and *Epeolus* are subtribally separated.

Nomada armata Herrich-Schaeffer

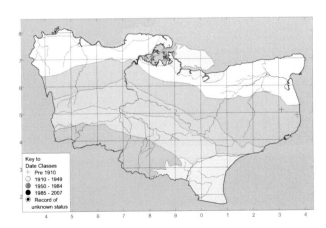

This nationally endangered cleptoparasitic bee has not been recorded in the county for over a century and is assumed extinct here. Should it reoccur, it will be where the host *Andrena hattorfiana* is found, on chalk grassland where *Knautia arvensis* is well established. This scabious is the principal forage plant of *A. hattorfiana*. *N. armata* is recorded visiting particularly *Knautia arvensis* but also *Scabiosa columbaria*, *Trifolium repens* and *Crepis vesicaria* in other counties.

Occurrence: 0, 1 *A Kent flight period cannot be defined*   1861, 1896
National status   Shirt: RDB1   Falk: pRDB1
Kent status   Present work: pKRDB1+

96 The Bees, Wasps and Ants of Kent

Nomada conjungens Herrich-Schaeffer

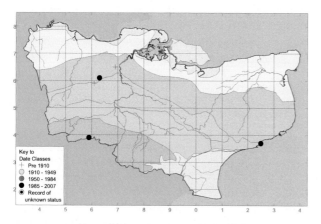

Recently refound in the county after an absence of records for a century. The host of this *Nomada* is *Andrena proxima*, itself an uncommon mining bee. There is a possibility that the *Nomada* is more widespread but over-looked, being found on the chalk scarps, Folkestone Warren and Tunbridge Wells sand. It has been recorded visiting *Smyrnium olusatrum*, *Oenanthe* and *Euphorbia* in other counties.

Occurrence: 3, 5   *v-vii*   1899-2004
National status   Shirt: RDB3   Falk: pRDB2
Kent status   Present work: pKRDB2

Nomada fabriciana (Linnaeus)

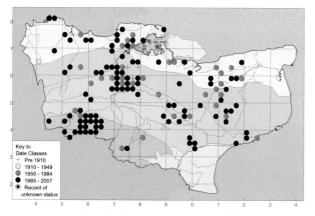

An abundant cleptoparasitic bee hosted by *Andrena bicolor*, *A. minutula* and probably other species in the *A. minutula* group. Rarely recorded on the coast and scarce on the weald clay but frequent elsewhere. Flower visits are recorded from *Taraxacum*, *Epilobium*, *Salix*, *Euphorbia* and *Alliaria petiolata*.

Occurrence: 97, 135   *iii-viii*   1892-2007
No status.

Nomada ferruginata (Linnaeus)
Synonymy: *N. xanthosticta*

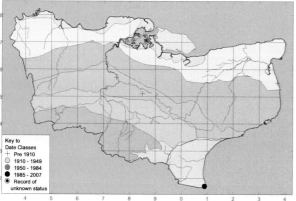

The host of this endangered cleptoparasite is *Andrena praecox*, a local mining bee. The only modern county record is from the Dungeness shingle but the bee was formerly known from lower greensand in the Maidstone area. Adult *N. ferruginata* have been recorded from *Salix* catkins but Asteraceae may also be visited.

Occurrence: 1, 5   *iv-vii*   1850s-1988
National status   Shirt: RDB1   Falk: pRDB1
Kent status   Waite: KRDB2   Present work: pKRDB1

Nomada flava Panzer

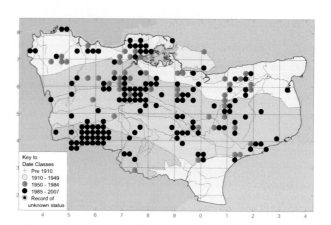

The apparent lack of records before 1953 is probably due to confusion of this species with *N. ruficornis* and *N. panzeri*. *N. flava* is an abundant cleptoparasite of *Andrena carantonica* (= *scotica*) and possibly some related *Andrena* spp.. The parasite is particularly frequent on the sands and chalk, but also found on clay soils. Adults have been found visiting *Prunus* (Rosaceae), *Acer*, *Euphorbia amygdaloides*, Apiaceae, Brassicaceae and *Taraxacum* (Asteraceae).

Occurrence: 140, 189   *iv-vii*   1953-2007
No status.

Nomada flavoguttata (Kirby)

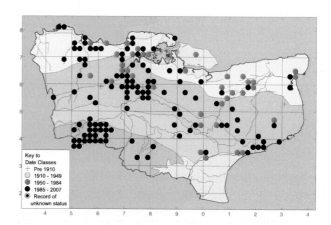

An abundant cleptoparasite of bees in the *Andrena minutula* complex. It is found on all soil types but is scarce on clays and rarely coastal. Adults have been recorded visiting *Prunus* and Asteraceae, including *Taraxacum* and *Senecio*.

Occurrence: 115, 156   *iii-ix*   1890s-2007
No status.

Nomada flavopicta (Kirby)

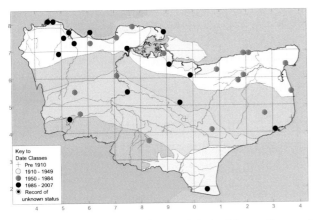

A widespread species across the county but not frequently recorded. Apparently absent from the weald clay and scarce on the chalk. A cleptoparasite of the mining bees, *Melitta leporina* and *M. tricincta*. Asteraceae are particularly recorded for flower visits.

Occurrence: 16, 36 *vi-ix* 1890s-2005
No status.

Nomada fucata Panzer

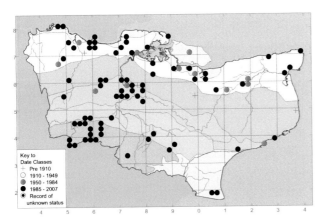

A twin brooded cleptoparasite that has commonly been recorded in the modern period. The host *Andrena flavipes* also has two broods. The *Nomada* apparently goes through cyclical periods of abundance which do not relate to the abundance of the host. The distribution of the parasite is positively correlated with sandy soils: sedimentary, alluvial and coastal. Flower visits are recorded for Asteraceae and *Prunus*; occasionally Apiaceae and Brassicaceae.

Occurrence: 73, 89 *iii-vi, vi-ix* 1850s-2007
National status Falk: pNa
Kent status Waite: Notable Present work: No Kent status.

Nomada fulvicornis Fabricius

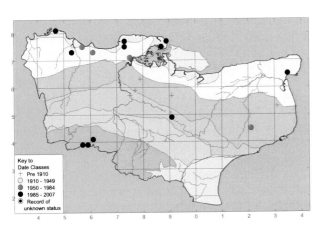

Kent is a stronghold for this bee, hosted by *Andrena bimaculata, A. tibialis* and *A. pilipes* (s.l.). The bee has two generations, as does some of the hosts. Many of the data are for the north Kent marshes but it is also found inland on sands and the chalk. I have no information for flower visits in the county but the first brood is reported from *Salix*, Brassicaceae, *Euphorbia, Armeria* and *Bellis*; the second from *Erica cineria, Rubus* and various Asteraceae.

Occurrence: 11, 21 *iv-vi, vii-ix* 1896-2007
National status Shirt: RDB3 Falk: pRDB3
Kent status Waite: No status Present work: pKa

Nomada goodeniana (Kirby)

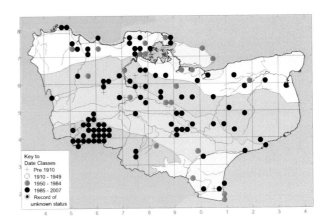

An abundant bee, hosted by several related *Andrena* species, including *A. nigroaenea* and *A. nitida*. As with the hosts, there is a single generation per year. The bee is found on all soil types but scarcest on clay soils. Flower visits are recorded for Asteraceae and *Alliaria petiolata* (Brassicaceae).

Occurrence: 92, 122 *iv-vii* 1893-2007
No status.

Nomada guttulata Schenck

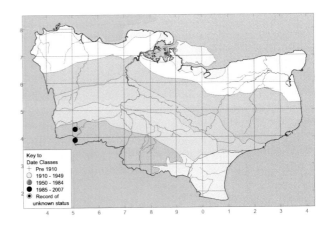

A very rare bee in the county with only three modern records. There are possible old data for "East Kent" but these have not been verified. This is a cleptoparasite of the mining bee *Andrena labiata*, itself a rarity in the county. Flower visits exist in the literature for *Veronica chamaedrys* (the host bee's food plant) and *Potentilla*.

Occurrence: 2, 2 *v* 2005-2007
National status Shirt: RDB1 Falk: pRDB1
Kent status Present work: pKRDB1

Nomada hirtipes Perez
Synonymy: *N. bucephalae*.

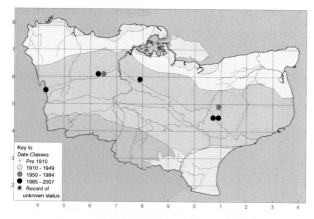

A rare cleptoparasite of solely *Andrena bucephala*. Although the host is found more widely, the parasite is recorded in Kent only from grassland on the chalk scarps. Literature data for flower visits include *Cardamine pratensis*, *Menyanthes* and *Prunus spinosa*.

Occurrence: 5, 7   iv-vi   1979-2002
National status   Shirt: RDB3   Falk: pRDB3
Kent status   Waite: KRDB2   Present work: pKRDB3

Nomada integra Brullé
Synonymy: Formerly misidentified as *N. pleurosticta*

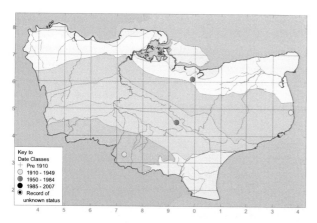

This is a cleptoparasitic bee that has shown a definite decline in recent decades, as indeed has the host, *Andrena humilis*. Most records are for sandy soils, apart from one near the coast. The parasite is recorded in other counties as visiting *Ranunculus bulbosus*, *Fragaria*, *Cerastium* and some Asteraceae.

Occurrence: 0, 6   v-vii   1897-1979
National status   Falk: pNa
Kent status   Present work: pKRDB1

Nomada lathburiana (Kirby)

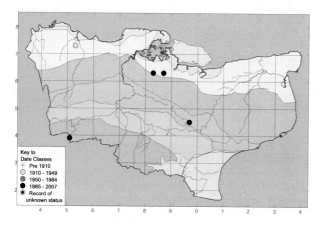

This cleptoparasitic bee is spreading with its host *Andrena cineraria*, and may become more common in Kent as a result. Most data are from on sandy soils but the Queendown Warren site is chalk grassland. In other counties *N. lathburiana* is recorded as visiting *Prunus*, *Salix*, *Ribes* and *Taraxacum*.

Occurrence: 4, 5   v   1918-2006
National status   Shirt: RDB3   Falk: pRDB3
Kent status   Present work: pKRDB3

Nomada leucophthalma (Kirby)

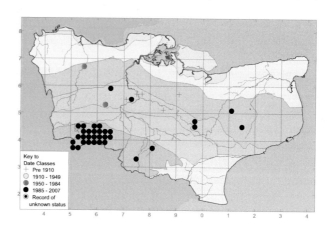

A widespread bee in the Tunbridge Wells area but infrequently recorded elsewhere in the county. This may be in part due to observer bias – the bee has an early flight period, as does its hosts, particularly *Andrena clarkella* and to a lesser extent, *A. apicata*. Most data are from on sandy soils with a few from the chalk. There is little information for flower visits but *Salix* is used, in common with its hosts.

Occurrence: 33, 41   iii-v   1893-2007
No status.

Nomada marshamella (Kirby)

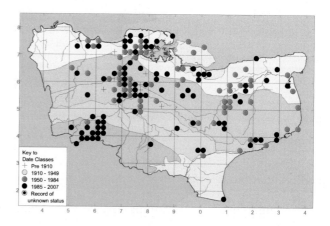

An abundant cleptoparasite with a bimodal phenology. The second peak in numbers is much smaller than the first, indicating that the second brood is less common than the first. The hosts are the *Andrena trimmerana* complex of bees, most often being the single brooded *A. carantonica* (= *scotica*) but probably also including *A. trimmerana* itself, though less frequently. The parasite visits a variety of plant families, including Rosaceae, Asteraceae, Apiaceae and Ericaceae.

Occurrence: 85, 157   iii-vii, vii-ix   1893-2007
No status.

Nomada panzeri Lepeletier
Synonymy: Once confused with *N. flava* under the name *N. ruficornis*.

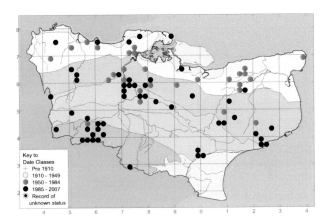

A common cleptoparasitic bee widely distributed across the county but rare in low lying wetlands. The lack of early data is due to confusion with *N. flava* and *N. ruficornis*. Most frequent on the chalk and sands. The hosts are mining bees of the *Andrena helvola* complex, including *A. helvola* and *A. varians*. It visits Brassicaceae, *Acer* and *Euphorbia*, and a female was recorded with unidentified orchid pollinia attached to the face.

Occurrence: 51, 79   *iv-vii*   1957-2007
No status.

Nomada ruficornis (Linnaeus)
Synonymy: *N. bifida*

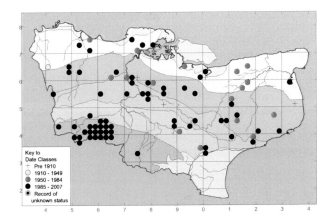

This bee had been the subject of confused identity until the work of R C L Perkins (1919). It is a very common cleptoparasite hosted solely by *Andrena haemorrhoa*. Like *N. panzeri*, it is rare in low lying areas but is relatively more frequent on clay soils. It is often found on the margins of woodland. The bee has been recorded visiting Rosaceae (*Crataegus* and *Prunus*) and *Senecio* (Asteraceae).

Occurrence: 64, 85   *iii-vi*   1890s-2007
No status.

Nomada rufipes Fabricius

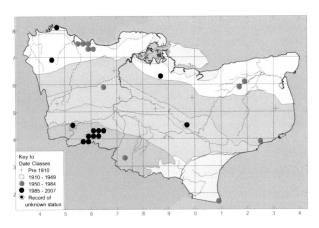

This is a widely distributed but not frequent cleptoparasite hosted by the *Andrena fuscipes* group. In Kent, the main hosts are *A. fuscipes* and *A. denticulata* but it will also parasitise *A. nigriceps* and *A. simillima*. There are a number of occurrences of the parasite where there are none of the named hosts, particularly on the north Kent marshes. This may be due to incomplete observations of the host species. I have no information on flower visits.

Occurrence: 13, 30   *vii-ix*   1896-2007
Kent status   Present work: pKb

Nomada sheppardana (Kirby)
Synonymy: *N. furva* (misident.)

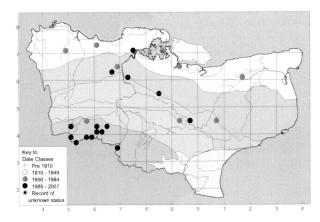

Although not classified, this is not a frequently recorded species apart from in the Tunbridge Wells area. It is a parasite of *Lasioglossum nitidiusculum* and particularly *L. parvulum*. Most data are for sandy soils but there are also some for the chalk scarps. Again, there are no flower visit data for Kent.

Occurrence: 15, 23   *iv-vii*   1963-2006
No status.

Nomada signata Jurine

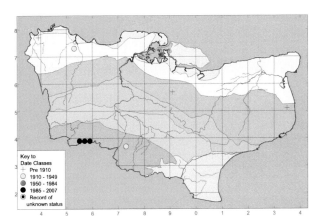

Although hosted by the frequent *Andrena fulva*, this is a very rare bee. It is not usually recorded from gardens in Kent, frequent habitat for *A. fulva*. Instead it prefers heathland and in the modern era, is only recorded from Tunbridge Wells. Flower visits in the UK are for *Salix*, *Taraxacum* and *Euphorbia*.

Occurrence: 3, 6 *iv-v, viii* 1895-2007
National status   Shirt: RDB3   Falk: pRDB2
Kent status   Waite: KRDB1   Present work: pKRDB2

Nomada striata Fabricius

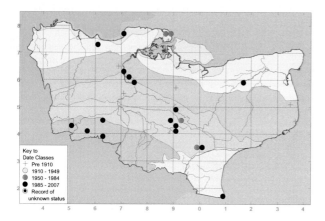

An infrequently recorded cleptoparasite of *Andrena wilkella* and *A. ovatula*. The parasite is distributed across all the main soil types in the county, including the weald clay. I have no flower visit data for the species in Kent.

Occurrence: 16, 27   *iii-vii*   1896-2006
No status.

Epeolus cruciger (Panzer)

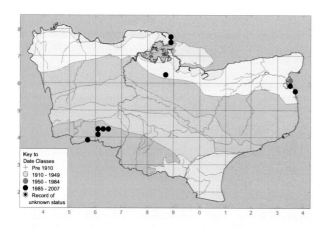

A rather scarce bee in the county, being most frequent on sandy heath in the UK. In Kent, is also found on coastal sands. This small bee is a cleptoparasite of *Colletes*, most frequently *C. succinctus* but also *C. marginatus* (and in Europe, *C. hederae*). The species is known to visit *Calluna* and *Epilobium*, but must visit many other flowers as well.

Occurrence: 10, 12   *vii-ix*   1850s-2007
Kent status   Present work: pKb

Epeolus variegatus (Linnaeus)

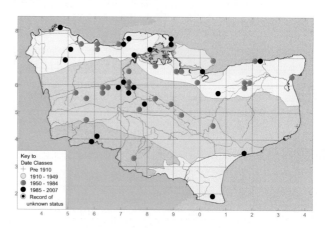

More frequent than the last species but still not common; possibly declining. The hosts are *Colletes daviesanus*, *C. fodiens*, *C. halophilus* and *C. similis*. The form on *C. halophilus* is a little larger than average. *E. variegatus* is found on sandy soils and coastal saltmarsh, rare otherwise. It frequently visits *Senecio jacobaea* (Asteraceae) and probably other flowers.

Occurrence: 21, 52   *vii-x*   1895-2007
No status.

Tribe Apini.

♀ *Anthophora bimaculata*

A large diverse group world wide, of which one species, *Apis mellifera*, has been transported by man across much of the globe and is now virtually cosmopolitan. The tribe can be broken into four British subtribes, as mentioned under the head of the subfamily. A discussion of the four subtribes follows:

Subtribe Eucerina: One of the smallest of the four with two British species, but many more in southern Europe. Both the British species, the long-horned bees, have been recorded from Kent but one is now probably extinct in the UK.

Subtribe Anthophorina: A subtribe with seven British species, one of which is probably extinct in the UK and another at least so in Kent. There are two genera, *Anthophora* and its cleptoparasite, *Melecta*. *Anthophora* tend to be ground nesters, although *A. furcata* probably nests in rotten tree stumps.

Subtribe Bombina: The first subtribe of British "corbiculate" bees. These are the bumblebees, with 24 of the 26 or so British species recorded in the county. All bumbles are now placed in *Bombus*, this genus including the social parasites as well as the free-living species. A species absent from Kent is *B. monticola*. It appears that *B. subterraneus* has become extinct in the county and the UK in recent decades. *B. pomorum* may have been a native Kent species at the extremity of its range but has been extinct for over a century. *B. cullumanus* once occurred in Kent but is also regarded as extinct in the UK. *B. magnus* and *B. cryptarum* may represent good species separate from *B. lucorum* and both could be recognised, *B. cryptarum* possibly being present in the Kent fauna. The introduced *B. hypnorum* is a recent addition to the Kent fauna.

Some species of *Bombus* can prove difficult to identify accurately, the smaller workers not always proving "good taxonomic material". Indeed even the larger workers of *B. terrestris* cannot always be separated from those of *B. lucorum*. It is often the case that males are most easily identified and capturing these probably does least harm to sensitive populations.

Bumbles are well known to be social insects with a female worker caste moderately well defined from the queens. In the advanced species there is a distinct size gap between the two castes but this is not always the case with less complex species. In a few, like *Bombus pascuorum*, the smallest queens are smaller than the largest workers and constitute more a physiological caste, shadowing the case in the social halictine species.

The socially parasitic *Bombus* have been well documented although sometimes important information on the life cycle is lacking, e.g. whether the female of a particular parasitic species usually kills the host queen or if the latter can be facultatively deposed as the alpha female in the colony. It appears than the parasite often does not really become part of the colony and in some species she may need the presence of the host queen's pheromones to prevent ovarian development in the workers. She may also be killed by the workers in the same way that the host queen sometimes is at the end of the colony life cycle.

Subtribe Apina: The second subtribe of corbiculate bees. This contains a single species in Britain, the hive bee or honeybee, *Apis mellifera*, which was introduced from the near continent, probably during the Roman period or earlier. Its distribution is mostly controlled by beekeepers although sometimes a swarm will set up a nest in a hollow tree and survive for several years. It is mapped here only for completeness sake.

Eucera longicornis (Linnaeus)

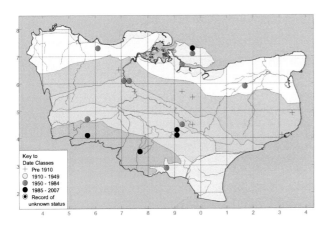

An interesting bee which has declined considerably in recent decades, from being regularly recorded in the late 1970s and early 1980s. The bee nests in aggregations in sparsely vegetated soils or short turf and provisions with leguminous pollen. Flower visits are known from several other plant families. This bee is a host of *Nomada sexfasciata*, an endangered cleptoparasitic bee not known from Kent.

Occurrence: 5, 21 *v-vii* 1879-2007
National status Falk: pNa
Kent status Waite: Notable Present work: pKRDB3

Eucera nigrescens Perez
Synonymy: *E. tuberculata*

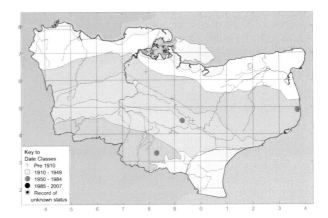

This bee is very similar to the last and probably became extinct in Britain only shortly after being recognised in our native fauna. Kent was its stronghold in the UK and despite much searching it has not been seen since 1970 (coll. J C Felton), although an unconfirmed record exists for 1975. Pollen is obtained from legumes as in the last species but the only known flower visit in the UK is for *Vicia cracca*. On the continent a host for *Nomada sexfasciata*.

Occurrence: 0, 9 *v-viii* 1901-1970
National status Shirt: RDB1 Falk: pRDB1
Kent status Waite: KRDB1 Present work: pKRDB1+

Anthophora bimaculata (Panzer)

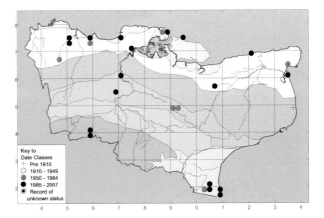

Although not classified, this is not a common species in the county, being much more frequent on sandy heaths. All Kent records are from the various sands or coastal shingle. This is a mining bee nesting in hard packed sand, such as on firm paths and in soft rock cliffs. It visits Lamiaceae and Asteraceae such as *Inula conyzae*, but also many other plants like Dipsacaceae, *Rubus* and *Erica*.

Occurrence: 18, 26 *vi-ix* 1890s-2007
No status.

102 The Bees, Wasps and Ants of Kent

Anthophora furcata (Panzer)

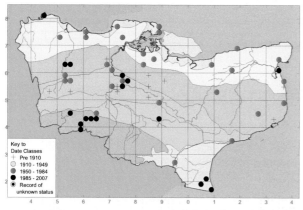

Not regarded here as common, and possibly declining in Kent. Widespread across the county but not often found on clay soils. It is believed that the female nests by gnawing into rotten tree stumps or utilising beetle exit holes in similar situations. Oligolectic on Lamiaceae, e.g. *Ballota nigra*.

Occurrence: 17, 57   *vi-viii*   1892-2007
No status.

Anthophora plumipes (Pallas)
Synonymy: *A. acervorum*, *A. pilipes*

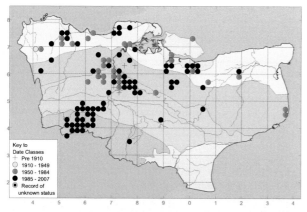

A very common bee frequently found in gardens. It often nests in the soft mortar of old walls, probably a substitute for soft rock cliff faces. Most frequently recorded on sandy soils but occasionally on chalk and the clays. It appears to be scarce in low-lying areas. The female visits mainly Lamiaceae and Boraginaceae for pollen, *Lamium* species are frequently observed. *Lathraea squamaria* (Orobanchaceae) is an interesting record. *A. plumipes* is the host of the scarce cleptoparasite *Melecta albifrons*.

Occurrence: 70, 105   *ii-vi*   1893-2007
No status.

Anthophora quadrimaculata (Panzer)

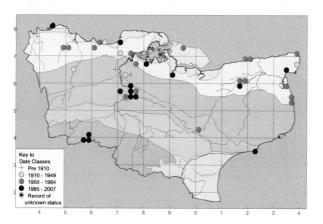

Kent appears to be a stronghold for this species in the UK, although it may be showing a decline even here. It is found on the sands and chalk but apparently not on clay soils. Like *A. plumipes*, it will nest in sand cliff faces and soft mortar in old walls. Pollen sources are unclear but the bee visits Lamiaceae such as *Nepeta*, *Lamium* and quite frequently cultivated *Lavandula*.

Occurrence: 15, 41   *v-ix*   1894-2007
National status   Falk: pNb
Kent status   Waite: Notable   Present work: pKb

Anthophora retusa (Linnaeus)

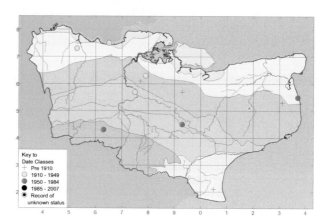

Despite repeated searches this bee has not been refound in the county, the last record being at Pluckley in 1966 (coll. J C Felton). Indeed, this is a nationally endangered species known from only a handful of modern sites in the UK. Modern data are from coastal soft rock cliffs. The bee was formerly recorded from across Kent on the sands and chalk, and the reason for its presumed extinction in the county is unknown. It is the host of the very rare cleptoparasite, *Melecta luctuosa*, itself probably extinct in the UK.

Occurrence: 0, 12   *v*   1898-1966
National status   Shirt: RDB3   Falk: pRDB1
Kent status   Present work: pKRDB1+

Melecta albifrons (Foerster)

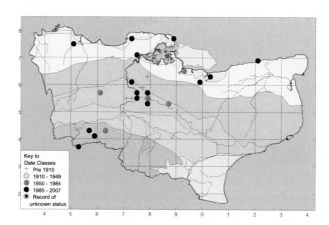

A rather uncommon cleptoparasite of *Anthophora plumipes*, the distributions of which tally quite closely. The *Melecta* can be found inspecting nest sites of *A. plumipes*. Flower visits for nectar only have been recorded for Lamiaceae, *Cardamine pratensis* and wallflowers.

Occurrence: 16, 24   *iii-vi*   1893-2007
Kent status   Present work: pKb

Melecta luctuosa (Scopoli)

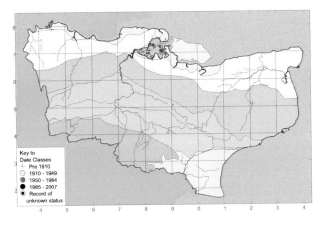

This bee is assumed long extinct in the county and indeed, in the UK. The last British record confirmed by a specimen was from the New Forest in 1912. British records were from heathland sands and soft rock cliffs, but Kentish data were principally from woodland on sandy soils. Two of the Kent localities are too vague to be mapped. The bee is a cleptoparasite of *Anthophora retusa*, itself probably extinct in Kent. Flower visits for nectar are for Boraginaceae and Lamiaceae.

Occurrence: 0, 3 A Kent flight period cannot be defined 1850s
National status Shirt: RDB1 Falk: pRDB1+
Kent status Present work: pKRDB1+

Worker *Bombus lucorum*

Bombus barbutellus (Kirby)

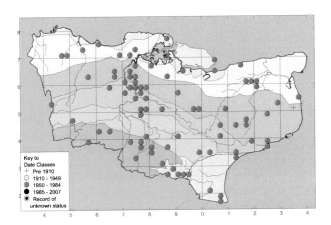

Not usually flagged up as being of conservation concern, this species is a cuckoo bumblebee that appears to have declined considerably since the 1970s and apparently has not been recorded since 1984. It could be misidentified and overlooked as its host, *B. hortorum*, still a common bee in the county. Flower visits for nectar are frequently to thistles but also teasel and *Rubus*.

Occurrence: 0, 84 iv-x 1892-1984
Kent status Present work: pKRDB2

Bombus bohemicus (Seidl)

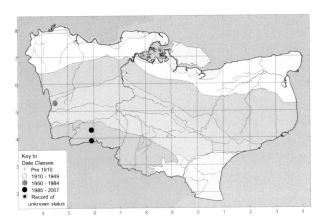

A very rare cuckoo bumble in the county, more typical of the north and west of the UK. Its host is the *B. lucorum* complex, of which the typical *B. lucorum* is plentiful in Kent. A male was recorded on *Rubus* and the bee is probably also found on thistles.

Occurrence: 2, 3 iv-ix 1972-2001
Kent status Present work: pKRDB2

Bombus campestris (Panzer)

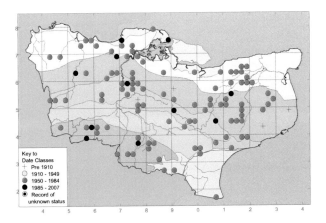

There are only five post 1989 records and the species has considerably declined. There are possible signs of recovery in other southern counties and the bee is plentiful in Pas-de-Calais, promising for a recovery in Kent. It is a cuckoo particularly of the abundant *Bombus pascuorum* and possibly other species of this group. The bee visits thistles and knapweeds for nectar but occasionally also other Asteraceae and *Rubus*. *Convolvulus* and "scabious" are also mentioned.

Occurrence: 11, 112 iv-ix 1890s-2007
Kent status Waite: Notable Present work: pKRDB3

Bombus cullumanus (Kirby)

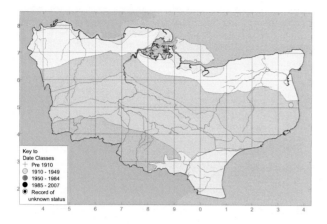

This free living bumblebee is in all probability extinct in the UK. It shows a strong preference for flower-rich calcareous grassland. There is only one old Kent record (coll. F W L Sladen), of a male visiting *Centaurea nigra*.

Occurrence: 0, 1   ix   1911
National status   Shirt: RDB1+   Falk: pRDB1+
Kent status   Present work: pKRDB1+

Bombus distinguendus Morawitz

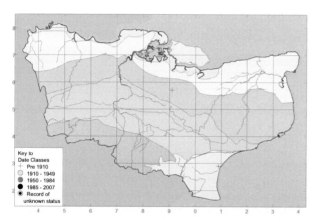

This bumblebee is considered extinct in Kent, although probably once native to the county. It is a large, conspicuous species that would not easily be over looked. In the UK it is now confined to the Western Isles and the extreme north of Scotland. It favours flower rich coastal grassland.

Occurrence: 0, 2   vii   1890s, 1900
National status   Falk: pNb
Kent status   Present work: pKRDB1+

Bombus hortorum (Linnaeus)

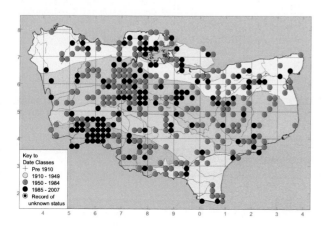

An abundant bumble found across the county but showing a possible slight decline. It appears to be slightly less common on the weald clay but this could be due to observer bias. The nest is usually constructed underground. The bee tends to favour flowers with a long corolla, such as *Digitalis* and Lamiaceae, but is widely polylectic. It is a host of the now rare *B. barbutellus*.

Occurrence: 124, 352   ii-x   1892-2007
No status.

Bombus humilis Illiger

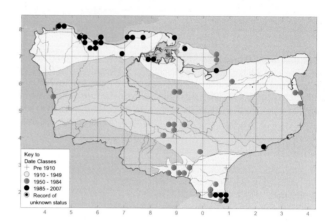

This is a bumblebee closely resembling the equally scarce *B. muscorum* and like that species, shows a modern retreat in distribution towards the coast. Inland records are now rare for this bee. It is a carder bumble and constructs its nest in grass tussocks which have had rodent nests present. The bee visits Lamiaceae, *Vicia*, *Centaurea*, thistles and occasionally *Digitalis*. A putative host for *B. campestris*.

Occurrence: 24, 54   v-x   1881-2007
Kent status   Waite: Provisionally notable   Present work: pKb

Bombus hypnorum (Linnaeus)

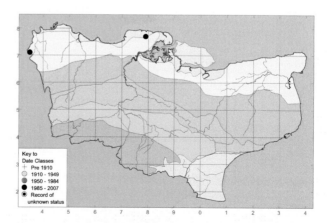

A very recent addition to the British fauna that seems to be spreading rapidly in the UK. It is likely to prove quite similar in habits to the related *B. pratorum*, which is placed in the same group, i.e. it has an early flight period and may become a garden species.

Occurrence: 2, 2   vi-viii   2005-2007
Kent status   Present work: pKRDBK

Bombus jonellus (Kirby)

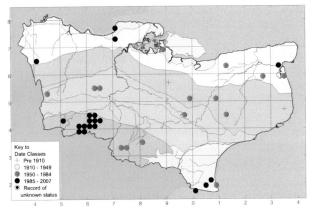

A further species in the *B. hypnorum* group but often described as an oligolege of Ericaceae. However, the queens fly in the spring long before native heathers are in flower and the species has been found in coastal areas of the county where there are no Ericaceae. It has also been found on characteristic heathand in the county, however. A worker was captured on *Aster tripolium* at Cliffe Marshes in September 2002. A possible host of *B. sylvestris*.

Occurrence: 19, 36   *iii-ix*   1890s-2007
Kent status   Present work: pKb

Bombus lapidarius (Linnaeus)

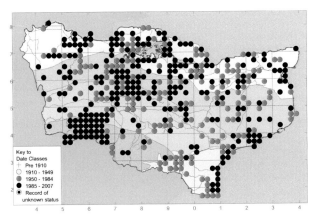

An abundant, free-living bumblebee and one of the most frequently recorded bees in Kent. Found generally across the county; the fewer data for the weald clay may be due to observer bias. The female castes are widely polylectic and the male is frequently recorded taking nectar on thistles. Host of the scarce social parasite, *B. rupestris*.

Occurrence: 264, 433   *ii-xi*   1894-2007
No status.

Bombus lucorum (Linnaeus)

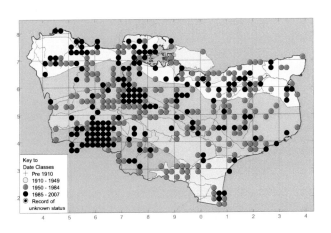

A very common bumblebee in which the worker caste is not always identifiable. This taxon may possibly include *B. cryptarum* in the county. Generally distributed across Kent. The female castes are widely polylectic and the male can frequently be found on thistles and *Rubus*. Host of the cuckoo bee, *B. bohemicus*, a rarity in Kent.

Occurrence: 161, 370   *i-xii*   1892-2007
No status.

Bombus muscorum (Linnaeus)

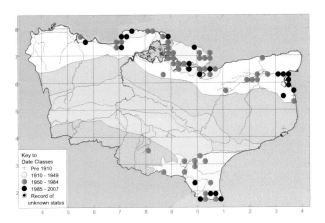

Closely related to *B. humilis*, this is another bumblebee retreating to the coast, where it still declines. It is a carder bee, nesting on the surface in grass tussocks. The females forage on Lamiaceae, Boraginaceae, *Rubus*, Asteraceae, *Trifolium*, *Lotus* and other plants. Males can be found on Asteraceae e.g. thistles, and *Rubus*, amongst other plants.

Occurrence: 23, 72   *iv-x*   1890s-2006
Kent status   Waite: Provisionally notable   Present work: pKb

Bombus pascuorum (Scopoli)

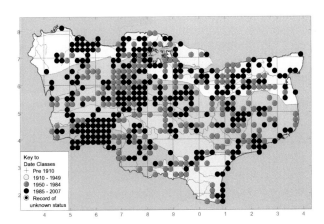

This is an abundant carder bumblebee found universally across the county. The female castes have a fondness for Lamiaceae but are widely polylectic. Males will also visit Asteraceae, including thistles, *Centaurea* and *Leucanthemum*, and also *Rubus*. The bee is the main host of the cuckoo bumble, *B. campestris*.

Occurrence: 296, 510   *ii-xi*   1893-2007
No status.

106 The Bees, Wasps and Ants of Kent

Bombus pomorum (Panzer)

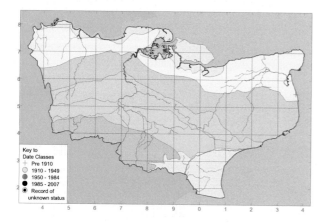

Only two 19th century records of this bee are known to me; a few males captured on the Deal sandhills and a queen in flight in June, so this could possibly be a case of an early extinction (Falk, 1991). The female castes have been found on *Trifolium pratense* and the males favour *Knautia*.

Occurrence: 0, 2 *vi* 1837
National status Shirt: Appendix   Falk: Appendix
Kent status Present work: pKRDB1+

♂ *Bombus pratorum*
© Lee Manning

Bombus pratorum (Linnaeus)

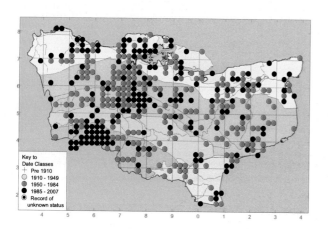

Clearly an abundant bumblebee that has an early flight period. The capture of workers quite late in the season indicates a second colony cycle in some years. Generally distributed across the county. The female castes are widely polylectic, foraging on *Salix*, Lamiaceae, *Prunus* and many other plants. The workers and males are often to be found on *Rubus*, including cultivated raspberries. Host of the cuckoo bumblebee, *B. sylvestris*.

Occurrence: 162, 393 *i-xi* 1894-2007
No status.

Bombus ruderarius (Mueller)

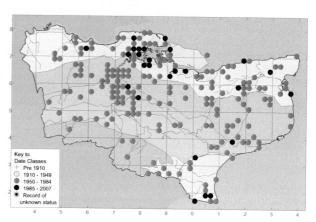

This carder bumblebee has also declined considerably in the last two decades and is given county notable status. Widely polylectic but favours Lamiaceae, *Vicia*, common comfrey and other flowers with fairly long corollas. A putative host for the scarce cuckoo bumble, *B. campestris*.

Occurrence: 26, 203 *iii-x* 1903-2007
Kent status Waite: Provisionally notable   Present work: pKb

Bombus ruderatus (Fabricius)

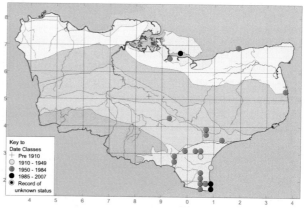

This is clearly a much declined bumblebee in Kent, although its recording is severely hampered by the close resemblance of the workers to those of *B. hortorum*. The queens, however, are more distinct and can often be identified in the field. The bee visits much the same flowers as the closely related *B. hortorum*: *Digitalis*, *Convolvulus*, *Ballota*, *Echium*, hollyhocks etc.

Occurrence: 3, 25 *iv-ix* 1882-2002
National status Falk: pNb
Kent status Waite: KRDB2   Present work: pKRDB2

Bombus rupestris (Fabricius)

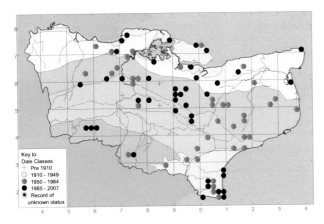

A cuckoo bumblebee that passed through a period of quite serious decline but which is showing signs of a recovery. The host is the abundant *B. lapidarius*. Males have been recorded from a variety of flowers, including thistles, *Centaurea*, *Lythrum* and *Rubus*.

Occurrence: 41, 86   *v-x*   1882-2007
National status   Falk: pNb
Kent status   Waite: Notable   Present work: pKb

Bombus soroeensis (Fabricius)

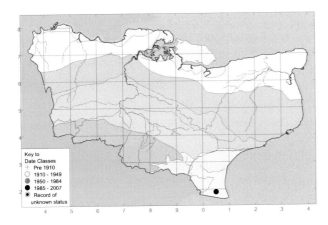

Although this is a very rare bumblebee in the county, it is known that there are further but undisclosed data for Kent. The bee is not frequent in south-east England in general and little is known about its habits here.

Occurrence: 1, 1   *viii*   2003
Kent status   Present work: pKRDBK

Bombus subterraneus (Linnaeus)

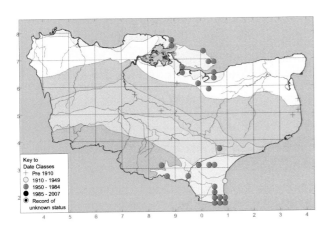

This large bumblebee serious declined in the late 1970s and is considered probably extinct in the UK. Kent was a stronghold for the species and it has not been recorded here since the 1980s in spite of targeted searches since. The bee was frequently recorded on *Ballota* (Lamiaceae) but also visited thistles, Dipsacaceae and *Digitalis*.

Occurrence: 0, 32   *vi-x*   1882-1980s
National status   Falk: pNa
Kent status   Waite: RDBX   Present work: pKRDB1+

Bombus sylvarum (Linnaeus)

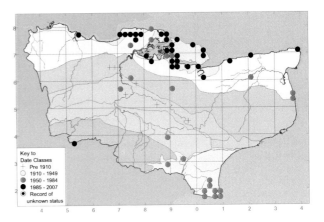

A very scarce carder bumblebee which seriously declined nationally in the 1980s. It was formerly present on the shingle at Dungeness but has not been recorded there since the 1980s, according to the known data. It has survived on brown field sites and grazing marsh in the Thames, Medway and Swale estuaries and in the last few years has shown possible signs of a recovery. For example, it has recently been recorded at Whitstable and Foreness, new sites for the species (coll. R I Moyse). The bee has been recorded visiting *Ballota*, *Digitalis*, thistles, *Centaurea* and *Lythrum*, and in recent years, *Epilobium hirsutum* and *Hirschfeldia incana*.

Occurrence: 31, 61   *iv-x*   1897-2007
National status   Falk: pNb
Kent status   Waite: KRDB2   Present work: pKRDBK

Bombus sylvestris (Lepeletier)

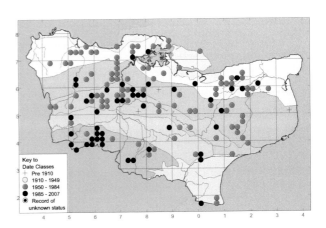

This is a common cuckoo bumblebee widely distributed across the county. It appears to be scarce on the weald clay. The main host is *B. pratorum* although other species in the *B. hypnorum* group may be used as alternatives. Like *B. pratorum* it may have two cycles per year. The bee has been recorded as visiting a variety of plants for nectar only, including *Salix*, Lamiaceae and dandelions, and the males can be frequent on *Rubus*, yellow flowered composites, thistles etc.

Occurrence: 46, 142   *iii-x*   1892-2007
No status.

Bombus terrestris (Linnaeus)

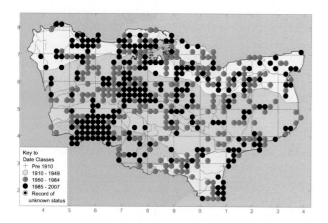

An abundant bumblebee generally distributed across the county. It may be slightly less frequent on the weald clay but this could also be due to observer bias. The bee is common in gardens and often nests in underground, abandoned rodent nests. It is widely polylectic, foraging on many different plant families. It is well known for the workers' habit of biting through the base of flowers with a long corolla to "rob" nectar and thus not effecting pollination. The species is known in southern Britain sometimes to found colonies in the autumn that over winter. *B. terrestris* is the host of the common cuckoo bumblebee, *B. vestalis*.

Occurrence: 250, 477   *i-xii*   1888-2007
No status.

♀ *Bombus vestalis*
© Lee Manning

Bombus vestalis (Geoffroy in Fourcroy)

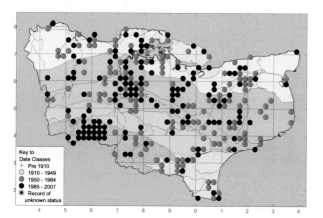

The most common cuckoo bumblebee in the county and an abundant species. The host is *B. terrestris*. The female bee visits a wide variety of plants for nectar only; Lamiaceae, *Crataegus*, *Glechoma*, ornamental Ericaceae, *Fragaria*, *Rubus*, *Hyacinthoides* and others. Males have recorded from thistles, *Centaurea*, *Pyracantha*, *Hebe*, *Eupatorium*, *Rubus*, *Lavandula*, *Lythrum* and many other flowers.

Occurrence: 141, 277   *ii-x*   1893-2007
No status.

Apis mellifera Linnaeus

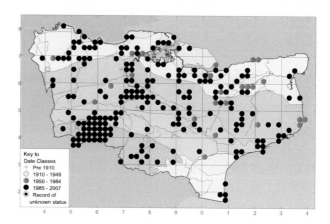

An abundant bee in the county, although the distribution map shows considerable observer bias. This is the only species of bee that is kept in hives in the UK. It is of extreme economic importance and is probably not native to Britain although thoroughly naturalised. It over winters as a viable colony clustered within the hive or hollow tree and workers can fly anytime it is mild enough, even in winter. It is widely polylectic and will forage on *Mahonia* in the depths of winter, if the temperature rises high enough for the cluster to loosen. Colony reproduction is by swarming and the queen never forages, workers always being present. Numbers of colonies in recent years may have been affected by the spread of the parasitic mite, *Varroa* and by the increase in the distribution of the bee wolf, *Philanthus triangulum*, a solitary wasp predator of honeybees.

Occurrence: 203, 232   *i-xi*   1897-2007
No status.

Appendix 1. Aculeate species possibly but inconclusively recorded from Kent

Myrmica hirsuta Elmes (Formicidae)
This species is of unknown status in the county. It is a social parasite (inquiline) of *M. sabuleti*. The sole record is of small winged female *Myrmica* taken near a *M. sabuleti* nest and possibly coming from it.

Myrmica vandeli Bondroit (Formicidae)
A worker was taken from a nest at Folkestone which strongly resembles *M. vandeli* but this requires confirmation. The species, only recently added to the British list, is a probable temporary social parasite of *M. scabrinodis*.

Formica rufibarbis Fabricius (Formicidae)
Occasionally, worker ants of the *F. fusca* group are found in the county that apparently key to *F. rufibarbis*. This is an endangered ant in the UK, reliably recorded only from a few Surrey heaths and the Isles of Scilly in modern times. It seems likely to me that the few known Kent specimens are just rather red, hairy *F. cunicularia*, particularly as only isolated individuals are found and in areas where *F. cunicularia* is common.

Evagetes dubius (Vander Linden) (Pompilidae)
It is published in the literature that this cleptoparasitic spider wasp occurs or has occurred in Kent. I have no data to confirm this.

Anoplius viaticus (Linnaeus) (Pompilidae)
There are three females without data labels in the Maidstone Museum collection that could be from the county. These specimens may possibly refer to the record of "*Pompilus bicolor*" from Upper Halling by H Lamb (VCH). Reliable modern data for the species are from sandy heaths in southern England and there are no further Kent records known to me.

Eumenes coarctatus (Linnaeus) (Vespidae)
This is the heath potter wasp, found locally in some other south-eastern counties of the UK, usually on *Calluna/Erica* heath. It is a distinctive wasp and although occasionally claimed from the county, I have been unable to confirm any specimens.

Ammophila pubescens Curtis (Sphecidae)
Known by the vernacular name of the "heath sand wasp", this species is occasionally claimed to have been recorded in Kent; e.g. in the VCH (1910) it is recorded from Wichling by Norton, under the name *A. campestris*. If this specimen was correctly identified, it probably represents a vagrant and I have seen no confirmed *A. pubescens* from Kent.

Lasioglosssum rufitarse (Zetterstedt) (Apidae)
There are two records from 1978 in the card index at Maidstone Museum, referred to this species. It is considered here that these are doubtful.

Appendix 2. Aculeate species that might yet be found in Kent

Hedychridium coriaceum (Dahlbom) (Chrysididae)
A *Hedychridium* was sighted at Bedgebury Forest near a *Lindenius albilabris* nesting aggregation but unfortunately was not captured. This may possibly have been *H. coriaceum*, a known parasite of *L. albilabris*. It was not seen on subsequent visits.

Lasioglossum cupromicans (Perez) (Apidae)
This bee has a mainly northern and western distribution in the UK but has apparently been recorded from Sussex and Essex. It may yet be found in Kent.

Appendix 3. Unusual and inconclusive aculeate specimens found in the county

Chrysis cf. ruddii Shuckard (Chrysididae)
A rubytailed wasp captured at Oaken Wood, Barming in 1998 looked unusual. It was determined as *C. impressa* by M Edwards but as *C. ruddii* by Dr M E Archer, which species it does closely resemble. However, the usual host of *C. ruddii*, *Ancistrocerus oviventris*, has not been found in the Maidstone area for nearly a century.

Andrena cf. semilaevis Perez (Apidae)
A male *Andrena minutula* group resembling the common *A. semilaevis* was captured at Senacre, Maidstone, in 1998. However, the puncturation of the mesonotum is much closer, and the apical impressions of the tergites are not perfectly smooth and shining.

Lasioglossum cf. fulvicorne (Kirby) (Apidae)
A female *Lasioglossum* close to *L. fulvicorne* was found visiting goldenrod in my garden at Senacre, Maidstone in 2003 but differs slightly in some respects. It is smaller than average for *L. fulvicorne* and the puncturation of the mesonotum is distinctly coarser and more widely spaced.

Nomada sp. (Apidae)
An unusual *Nomada* was captured at Oaken Wood, Barming in 2004. This male keys to the *N. flava/panzeri* couplet of Perkins (1919) but is very dark in colour. It does not key cleanly out of a draft key to *Nomada* by Else and its identity is unknown.

Natural England Natural Areas of Kent

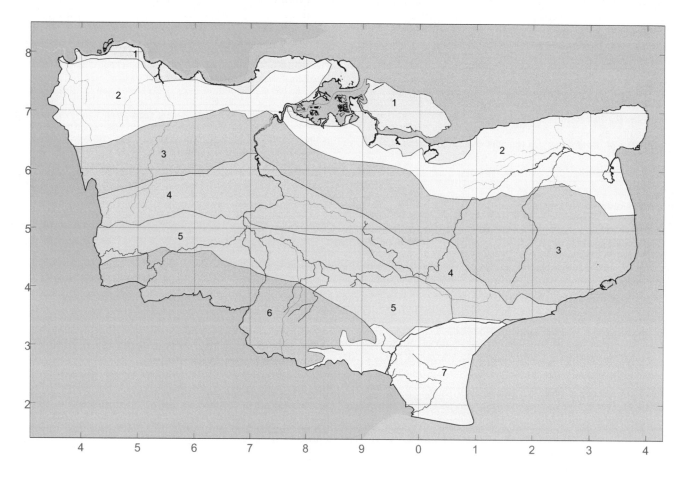

The government nature conservation body, Natural England, has subdivided the country into a series of Natural Areas, each of which has a characteristic association of geology, wildlife, land form and land use. They are intended to provide a wider context for the targeting of nature conservation action, and are used here to provide an indication of the relationship between the distribution of species and the features of the landscape.

More detailed information on each of the Natural Areas occurring in Kent can be found on the Natural England website.

1 Greater Thames Estuary

2 North Kent Plain

3 North Downs

4 Weald Greensand (Lower Greensand)

5 Low Weald & Pevensey

6 High Weald

7 Romney Marshes

References & Further Reading

Allen, G.W. (1982). Records of two uncommon Hymenoptera in Kent. *The Transactions of the Kent Field Club* **9**; Part 2.

Allen, G.W. (2006). Hymenoptera Aculeata Report 2005. *The Bulletin of the Kent Field Club: Number* **51**.

Bohart, R.M. & Menke, A.S. (1976). *Sphecid Wasps of the World, a generic revision.* University of California Press, California.

Bolton, B. (1994). *Identification Guide to the Ant Genera of the World.* Harvard University Press, Cambridge, Massachusetts.

Brothers, D.J. (1975). Phylogeny and classification of the aculeate Hymenoptera, with special reference to Mutillidae. *The University of Kansas Science Bulletin* **50**: 483-648.

Collingwood, C.A. (1958). The ants of the genus *Myrmica* in Britain. *Journ. Soc. Brit. Ent.*, **13**: 69-96.

Day M.C. (1988). *Spider Wasps. Hymenoptera Pompilidae.* Handbooks for the Identification of British Insects, **6**, Part 4. Royal Entomological Society.

Else, G.R. (2005). Section 10 – Check List of British Hymenoptera Aculeata. *BWARS Members' Handbook.*

Else, G.R. & Felton, J.C. (1994). *Mimumesa unicolor* (Vander Linden, 1829) (Hymenoptera: Sphecidae) a wasp new to the British list, with observations on related species. *Entomologist's Gazette* **45**: 107-114.

Falk, S.J. (1991). *A review of the scarce and threatened bees, wasps and ants in Great Britain.* Research and Survey in Nature Conservation. No 35. Peterborough: Nature Conservancy Council.

Felton, J.C. (1967). A Preliminary Account of the Ants (Hymenoptera, Formicidae) of Kent. *The Transactions of the Kent Field Club,* **3**, Part 2.

Felton, J.C. (1969). Further Records of Ants (Hymenoptera, Formicidae) from Kent. *The Transactions of the Kent Field Club,* **3**, Part 3.

Felton, J.C. (1988). The genus *Trypoxylon* Latreille, 1809 (Hym., Sphecidae) in Kent, with a first British record for *T. minus* de Beaumont, 1945. *Entomologists Monthly Magazine* **124**: 221-224.

Gauld, I. & Bolton, B. [Eds.] (1988, 2nd Ed. 1996). *The Hymenoptera.* The Natural History Museum, London. Oxford University Press.

Gayubo, S.F. & Felton, J.C. (2000). The European species of the genus *Nitela* Latreille, 1809 (Hymenoptera: Sphecidae). *Annals de la Société Entomologique de France* **36**: 291-313.

Guichard, K.M. (2002). *Passaloecus turionum* Dalhbom, 1845 (Hymenoptera: Sphecidae) new to the British list. *Entomologist's Gazette* **53**: 33-36.

Melo, G.A.R. (1999). Phylogenetic relationships and classification of the major lineages of Apoidea (Hymenoptera) with emphasis on the crabronid wasps. *Natural History Museum, University of Kansas. Scientific Papers* Number 14: 1-55.

Morgan, D. (1984). *Cuckoo wasps (Hymenoptera, Chrysididae).* Handbooks for the Identification of British Insects, **6**, Part 5. Royal Entomological Society.

Perkins, R.C.L. (1919). The British species of *Andrena* and *Nomada*. *Transactions of the Entomological Society of London* (1919): 218-319.

Pulawski, W.J. (1984). The status of *Trypoxylon figulus* (Linnaeus, 1758), *medium* de Beaumont, 1945 and *minus* de Beaumont, 1945 (Hymenoptera: Sphecidae). *Proceedings of the California Academy of Sciences* **43**: 123-140.

Richards, O.W. (1980). *Scolioidea, Vespoidea and Sphecoidea (Hymenoptera, Aculeata).* Handbooks for the Identification of British Insects, **6**, Part 3(b). Royal Entomological Society.

Shirt, D. [Ed.] (1987). *Insects.* British Red Data Books, 2. Peterborough: Nature Conservancy Council.

Waite, A. [Ed.] (2000). *The Kent Red Data Book. A Provisional Guide to the Rare and Threatened Flora and Fauna of Kent.* Kent County Council.

Yarrow, I.H.H. (1954). *Psenulus schencki* Tournier, a psenine wasp new to the British List. *Journ. Soc. Brit. Ent.,* **5**: 41-42.

Yarrow, I.H.H. (1970). Some nomenclatorial problems in the genus *Passaloecus* Shuckard and two species not before recognised as British (Hym. Sphecidae). *Entomologist's Gazette* **21**:167-189.

Index of Covered Taxa

A
Ageniellini, 23.
Agenioideus cinctellus, **27**.
Agenioideus sericeus, **27**.
[Ammophila pubescens, 36].
Ammophila sabulosa, **36**.
Ammophilini, 36.
Ancistrocerus, 11.
Ancistrocerus antilope, 11, **31**.
Ancistrocerus gazella, 11, 30, **31**, 32.
Ancistrocerus nigricornis, **31**.
Ancistrocerus oviventris, 2, 11, 12, 30, **32**.
Ancistrocerus parietinus, 11, **32**.
Ancistrocerus parietum, 11, **32**.
Ancistrocerus scoticus, 12, **32**.
Ancistrocerus trifasciatus, 10, 11, **32**.
Andrena, 2, 3, 57, 62, 63, 76, 95, 96.
Andrena alfkenella, **63**.
Andrena angustior, **63**.
Andrena apicata, **63**, 98.
Andrena barbilabris, **63**, 84.
Andrena bicolor, **64**, 96.
Andrena bimaculata, **64**, 97.
Andrena bucephala, **64**, 98.
Andrena carantonica, 2, 4, **64**, 96, 98.
Andrena chrysosceles, **65**.
Andrena cineraria, **65**, 98.
Andrena clarkella, **65**, 98.
Andrena coitana, **65**.
Andrena denticulata, **66**, 99.
Andrena dorsata, **66**, 84.
Andrena falsifica, **66**.
Andrena ferox, **66**.
Andrena flavipes, **66**, 97.
Andrena fucata, **67**.
Andrena fulva, **67**, 100.
Andrena fulvago, **67**.
Andrena fuscipes, **67**, 99.
Andrena gravida, **68**.
Andrena haemorrhoa, **68**, 99.
Andrena hattorfiana, **68**, 95.
Andrena helvola, **68**, 99.
Andrena humilis, **68**, 98.
Andrena labialis, **69**.
Andrena labiata, **69**, 97.
Andrena lapponica, **69**.
Andrena marginata, **69**, 95.
Andrena minutula, 63, **69**, 96.
Andrena minutuloides, **70**.
Andrena nana, **70**.
Andrena nigriceps, **70**, 99.
Andrena nigroaenea, 4, **70**, 97.
Andrena nigrospina, see *A. pilipes* aggregate.
Andrena nitida, **71**, 97.
Andrena niveata, **71**.
Andrena ovatula, **71**.
Andrena pilipes aggregate, 63, **71**, 97.
Andrena polita, **71**.
Andrena praecox, **72**, 96.
Andrena proxima, **72**, 96.
Andrena rosae, **72**, 73.
Andrena scotica, see *A. carantonica.*
Andrena semilaevis, **72**.
Andrena similis, **72**.
Andrena simillima, **73**, 99.
Andrena stragulata, **73**.
Andrena subopaca, **73**.
Andrena synadelpha, **73**.
Andrena tarsata, **73**.

Andrena thoracica, **74**.
Andrena tibialis, **74**, 97.
Andrena trimmerana, 4, 63, **74**, 98.
Andrena vaga, **74**.
Andrena varians, **74**, 99.
Andrena wilkella, **75**, 100.
Andreninae, 2, 58, **62**.
Andrenini, 63.
Anergates atratulus, 16, **18**.
Anoplius concinnus, **28**.
Anoplius infuscatus, **29**.
Anoplius nigerrimus, **29**.
[Anoplius viaticus, 26].
Anthidium, 3, 87.
Anthidium manicatum, **87**, 88.
Anthophora, 3, 87, 88.
Anthophora bimaculata 94, 100, **101**.
Anthophora furcata, **102**.
Anthophora plumipes, **102**.
Anthophora quadrimaculata, **102**.
Anthophora retusa, **102**, 103.
Apis mellifera, 3, 58, 100, 101, **108**.
APIDAE, 2, 37, **58**.
Apinae, 58, **94**.
Apini, 100.
APOIDEA, 2, 36.
Aporus unicolor, **29**.
Arachnospila, 3, **26**.
Arachnospila anceps, 26, **27**, 28.
Arachnospila consobrina, **27**.
Arachnospila minutula, **27**.
Arachnospila spissa, **28**.
Arachnospila trivialis, **28**.
Arachnospila wesmaeli, **28**.
Argogorytes, 54, 55.
Argogortyes fargei, **56**.
Argogorytes mystaceus, **56**.
Astata boops, 9, **37**.
Astatinae, 37.
Auplopus carbonarius, **23**.

B
Bembicinae, 54.
[Bembix rostrata, 54].
BETHYLIDAE, 7.
Bombus, 3, 14, 58, 100, 101.
Bombus barbutellus, **103**, 104.
Bombus bohemicus, **103**, 105.
Bombus campestris, **103**, 104, 105, 106.
[Bombus cryptarum, 101, 105].
Bombus cullumanus, **104**.
Bombus distinguendus, **104**.
Bombus hortorum, 103, **104**, 106.
Bombus humilis, **104**, 105.
Bombus hypnorum, 101, **104**.
Bombus jonellus, **105**.
Bombus lapidarius, **105**, 107.
Bombus lucorum, 101, 103, **105**.
[Bombus magnus, 101]
[Bombus monticola, 100]
Bombus muscorum, 14, **105**.
Bombus pascuorum 101, 103, **105**.
Bombus pomorum, 101, **106**.
Bombus pratorum, 104, **106**.
Bombus ruderarius, **106**.
Bombus ruderatus, **106**.
Bombus rupestris, 105, **107**.
Bombus soroeensis, **107**.
Bombus subterraneus, 100, **107**.

Bombus sylvarum, **107**.
Bombus sylvestris, 105, 106, **107**.
Bombus terrestris, 101, **108**.
Bombus vestalis, **108**.

C
Caliadurgus fasciatellus, **23**.
Ceratina cyanea, 94, **95**.
Cercerini, 56.
Cerceris, 7, **56**.
Cerceris arenaria, **56**.
Cerceris quadricincta, **57**.
Cerceris quinquefasciata, **57**.
Cerceris ruficornis, **57**.
Cerceris rybyensis, **57**.
Cerceris sabulosa, **57**.
Ceropales, 23.
Ceropales maculata, **30**.
Ceropalinae, 23, 29.
Chelostoma campanularum, **89**.
Chelostoma florisomne, 10, 15, **89**.
CHRYSIDIDAE, 5, 7.
Chrysidinae, 7, **9**.
CHRYSIDOIDEA, 7.
Chrysis angustula, **10**, 31, 32.
Chrysis fulgida, **10**, 33.
Chrysis gracillima, **10**, 31, 39.
Chrysis ignita aggregate, 7, 9, **10**.
Chrysis ignita (sensu stricto), **10**, 31, 32.
Chrysis illigeri, **11**, 38.
Chrysis impressa, **11**, 31, 32.
Chrysis longula, **11**, 31, 32.
Chrysis mediata, **11**.
[*Chrysis pseudobrevitarsis*, 31].
Chrysis ruddii, **11**, 30.
Chrysis rutiliventris, **12**, 32.
Chrysis schencki, **12**.
Chrysis viridula, **12**, 30.
Chrysura radians, 9, **12**, 87.
Cleptes nitidulus, **13**.
Cleptes semiauratus, 12, **13**.
Cleptinae, 12.
Coelioxys, 3, 87, **91**.
Coelioxys conoidea, 92, **93**.
Coelioxys elongata, **93**.
Coelioxys inermis, **93**.
Coelioxys mandibularis, 3, **93**.
Coelioxys quadridentata, 93, **94**.
Coelioxys rufescens, 93, **94**.
Colletes, 2, 3, 58, 95, 100.
Colletes daviesanus, 58, **59**, 100.
Colletes fodiens, 58, **59**, 100.
Colletes halophilus, **59**, 100.
Colletes hederae, 58, **59**.
Colletes marginatus, **59**, 100.
Colletes similis, **60**, 100.
Colletes succinctus, **60**, 100.
Colletinae, 2, **58**.
Colletini, 58.
Crabro, 40.
Crabro cribrarius, **40**.
Crabro peltarius, **41**.
[*Crabro scutellatus*, 40].
CRABRONIDAE, 2, 23, **37**.
Crabroninae, 37.
Crabronini, 37, **40**.
Crossocerus, 2, 14, 40.
Crossocerus annulipes, **41**.
Crossocerus binotatus, **41**.
Crossocerus capitosus, **41**.
Crossocerus cetratus, **41**.

Crossocerus dimidiatus, **41**.
Crossocerus distinguendus, **42**.
Crossocerus elongatulus, **42**.
Crossocerus exiguus, **42**.
Crossocerus megacephalus, **42**.
Crossocerus nigritus, **42**.
Crossocerus ovalis, **43**.
Crossocerus palmipes, **43**.
Crossocerus podagricus, **43**.
Crossocerus pusillus, **43**.
Crossocerus quadrimaculatus, **43**.
Crossocerus styrius, **43**.
Crossocerus tarsatus, **44**.
Crossocerus vagabundus, **44**.
Crossocerus walkeri, **44**.
Crossocerus wesmaeli, **44**.
Cryptocheilus notatus, **24**.

D
Dasypoda, 85.
Dasypoda hirtipes, 85, **87**.
Dasypodaini, 86.
Didineis, 54.
Didineis lunicornis, **54**.
Dienoplus, see *Harpactus*.
Diodontus, 48, 50.
Diodontus insidiosus, **52**.
Diodontus luperus, **52**.
Diodontus minutus, **52**.
Diodontus tristis, **52**.
Dipogon subintermedius, **23**.
Dipogon variegatus, **23**.
Dolichoderinae, 19.
Dolichovespula media, **34**.
Dolichovespula norwegica, **34**.
Dolichovespula saxonica, 34, **35**.
Dolichovespula sylvestris, **35**.
DRYINIDAE, 7.
Dryudella pinguis, 9, **37**.

E
Ectemnius, 40.
Ectemnius cavifrons, **44**.
Ectemnius cephalotes, **45**.
Ectemnius continuus, **45**.
Ectemnius dives, **45**.
Ectemnius lapidarius, **45**.
Ectemnius lituratus, **45**.
Ectemnius rubicola, **45**.
Ectemnius ruficornis, **46**.
Ectemnius sexcinctus, **46**.
Elampinae, 7.
Elampus panzeri, 8.
EMBOLEMIDAE, 7.
Entomognathus, 40.
Entomognathus brevis, **46**.
Epeolus, 3, 58, 95.
Epeolus cruciger, 59, 60, **100**.
Epeolus variegatus, 59, 60, 95, **100**.
Episyron rufipes, **28**.
Eucera, 95.
Eucera longicornis, **101**.
Eucera nigrescens, **101**.
[*Eumenes coarctatus*, 30].
Eumeninae, 30.
Evagetes, 23.
Evagetes crassicornis, **28**.
[*Evagetes dubius*, 26].
Evagetes pectinipes, **28**.

F
Formica, 4, 15, 19.
Formica cunicularia, **20**.
Formica fusca, **20**.
Formica rufa, 17, **20**.
Formica sanguinea, 4, 17, **20**.
FORMICIDAE, 13, **15**.
Formicinae, **20**.
Formicini, 20.
Formicoxenini, 17.
Formicoxenus nitidulus, 4, 16, **17**.

G
Gorytes, 54.
Gorytes laticinctus, **55**.
Gorytes quadrifasciatus, **55**.
Gymnomerus laevipes, **31**.

H
Halictinae, 58, **75**.
Halictini, 76.
Halictus, 3, 57, 75, 76.
Halictus confusus, **76**.
Halictus eurygnathus, **76**.
Halictus maculatus, **76**.
Halictus rubicundus, **76**, 83, 84.
Halictus tumulorum, **77**.
Harpactus, **54**.
Harpactus tumidus, 9, 54, **55**.
Hedychridium ardens, 7, **8**.
[*Hedychridium coriaceum*, 46]
Hedychridium cupreum, **9**, 37.
Hedychridium roseum, **9**, 37, 56.
Hedychrum, 7.
Hedychrum niemelai, 7, **9**, 56, 57.
[*Hedychrum rutilans*, 7, 58].
Heriades truncorum, 3, **88**.
Hoplitis, 89.
Hoplitis (Anthocopa), see *Osmia spinulosa*.
Hoplitis claviventris, 87, 88, **91**.
Hoplitis spinulosa, see *Osmia spinulosa*.
Hylaeini, 58, **60**.
Hylaeus, 2, 57, 58, 60.
Hylaeus annularis, **60**.
Hylaeus brevicornis, **60**.
Hylaeus communis, **61**.
Hylaeus confusus, **61**.
Hylaeus cornutus, 60, **61**.
Hylaeus gibbus, **61**.
Hylaeus hyalinatus, 60, **61**.
Hylaeus pictipes, **62**.
Hylaeus punctulatissimus, **62**.
Hylaeus signatus, 60, **62**.
Hylaeus spilotus, **62**.
Hypoponera punctatissima, **15**.

L
Larrinae, see Crabroninae.
Larrini, 37, 38.
Lasiini, 21.
Lasioglossum, 3, 4, 57, 75, 76, 83, 95.
Lasioglossum albipes, **77**.
Lasioglossum calceatum, **77**.
[*Lasioglossum cupromicans*, 75].
Lasioglossum fratellum, **77**, 83.
Lasioglossum fulvicorne, **77**, 82, 83.
Lasioglossum laevigatum, **78**.
Lasioglossum lativentre, **78**, 82, 84.
Lasioglossum leucopus, **78**.
Lasioglossum leucozonium, **78**, 82.
Lasioglossum malachurum, 4, 76, 78, **79**, 84.
Lasioglossum minutissimum, **79**, 83.
Lasioglossum morio, **79**, 84.
Lasioglossum nitidiusculum, **79**, 82, 83, 99.
Lasioglossum parvulum, **79**, 82, 83, 99.
Lasioglossum pauperatum, **80**.
Lasioglossum pauxillum, **80**.
[*Lasioglossum prasinum*, 75].
Lasioglossum punctatissimum, **80**.
Lasioglossum puncticolle, **80**.
Lasioglossum quadrinotatum, **81**.
Lasioglossum semilucens, **81**.
Lasioglossum smeathmanellum, **81**, 83.
Lasioglossum villosulum, **81**.
Lasioglossum xanthopus, **81**, 85.
Lasioglossum zonulum, **82**, 85.
Lasius, 19.
Lasius alienus aggregate, **21**.
Lasius brunneus, **21**.
Lasius flavus, **21**.
Lasius fuliginosus, **21**.
Lasius meridionalis, **21**, 22.
Lasius mixtus, **22**.
Lasius niger aggregate, 21, **22**.
Lasius umbratus, **22**.
Lasius umbratus aggregate, 21, **22**.
Leptothoracini, see Formicoxenini.
Leptothorax, 16.
Leptothorax acervorum, 16, **17**.
Lestiphorus, 54.
Lestiphorus bicinctus, **55**.
Lindenius, 40.
Lindenius albilabris, **46**.
Lindenius panzeri, **46**.

M
Macropis, 85.
Macropis europaea, 85, **86**.
Megachile, 15, 87, 94.
Megachile centuncularis, **91**, 93, 94.
Megachile circumcincta, **92**.
Megachile dorsalis, 87, **92**, 93.
Megachile ligniseca, **92**.
Megachile maritima, 97, **92**, 93.
[*Megachile parietina*, 87]
Megachile versicolor, **92**.
Megachile willughbiella, 97, **93**, 94.
Megachilinae, 2, 9, 58, **87**.
Megachilini, 87.
Melecta, 105.
Melecta albifrons, **102**.
Melecta luctuosa, 102, **103**.
Melitta, 95.
[*Melitta dimidiata*, 85]
Melitta haemorrhoidalis, 85, **86**.
Melitta leporina, 85, **86**, 97.
Melitta tricincta, 85, **86**, 97.
Melittinae, 2, 58, **85**.
Melittini, 85.
Mellinini, 37.
[*Mellinus crabroneus*, 37].
Mellinus arvensis, 37.
[*Methocha articulata*, 13].
Microdynerus exilis, 10, **31**.
Mimesa, 8.
Mimesa bicolor, **48**.
Mimesa bruxellensis, **48**.
Mimesa equestris, **48**.
Mimesa lutaria, **48**.
Mimumesa, 48.
Mimumesa dahlbomi, **49**.
Mimumesa spooneri, **49**.

Mimumesa unicolor, **49**.
Miscophini, 37, **38**.
Miscophus, 23.
Miscophus ater, **38**.
Miscophus concolor, **38**.
Monosapyga clavicornis, **15**, 89.
Mutilla europaea, **14**.
MUTILLIDAE, 13, **14**.
Mutillinae, **14**.
Myrmecina graminicola, 16, **19**.
Myrmecinini, 19.
Myrmica, 16.
Myrmica lobicornis, **16**.
Myrmica rubra, **16**.
Myrmica ruginodis, **16**.
Myrmica sabuleti, **16**, 19.
Myrmica scabrinodis, **17**.
Myrmica schencki, **17**.
Myrmica specioides, **17**.
Myrmicinae, **16**.
Myrmicini, 16.
Myrmosa atra, **14**.
Myrmosinae, **14**.

N
Nitela borealis, **39**.
Nitela lucens, **39**.
Nomada, 3, 4, 64, 87, 95.
Nomada argentata, 69, **95**.
Nomada armata, 68, **95**.
Nomada conjungens, 72, **96**.
Nomada fabriciana, 64, 69, **96**.
Nomada ferruginata, 72, **96**.
Nomada flava, 4, 64, 66, **96**, 99.
Nomada flavoguttata, 69, 72, 73, **96**.
Nomada flavopicta, 86, **97**.
Nomada fucata, 66, **97**.
Nomada fulvicornis, 64, 71, 74, **97**.
Nomada goodeniana, 70, 71, 74, **97**.
Nomada guttulata, 69, **97**.
Nomada hirtipes, 64, **98**.
Nomada integra, 68, **98**.
Nomada lathburiana, 65, **98**.
Nomada leucophthalma, 63, 65, **98**.
Nomada marshamella, 4, 64, 66, 74, **98**.
[*Nomada obtusifrons*, 65].
Nomada panzeri 68, 74, 96, **99**.
[*Nomada roberjeotiana*, 73].
Nomada ruficornis, 68, 96, **99**.
Nomada rufipes, 67, **99**.
[*Nomada sexfasciata*, 95, 101].
Nomada sheppardana, 79, **99**.
Nomada signata, 67, **99**.
Nomada striata, 75, **100**.
Nomadini, 94, **95**.
Nysson, 54.
Nysson dimidiatus **54**, 56.
[*Nysson interruptus*, 54, 55, 56].
Nysson spinosus, **54**, 55, 56.
Nysson trimaculatus, **55**.
Nyssoninae, see Bembicinae.

O
Odynerus, 30
Odynerus melanocephalus, 12, **30**.
Odynerus spinipes, 11, 12, **30**.
Omalus, 7.
Omalus aeneus, **8**.
Omalus puncticollis, **8**.
Osmia, 2, 3, 15, 87, 89.
Osmia aurulenta, **89**.

Osmia bicolor, **89**.
Osmia caerulescens, 88, **90**.
Osmia leaiana, 2, 88, **90**.
Osmia niveata, **90**.
Osmia pilicornis, 89, **90**.
Osmia rufa, 2, 87, **90**.
Osmia spinulosa, 87, **91**.
Osmia xanthomelana, **91**.
Oxybelini, 37, **47**.
Oxybelus argentatus, **47**.
Oxybelus uniglumis, **47**.

P
Panurgini, 75.
Panurgus, 62.
Panurgus banksianus, **75**.
Panurgus calcaratus, **75**.
Passaloecus, 8, 48, 50.
Passaloecus clypealis, **52**.
Passaloecus corniger, **53**.
Passaloecus eremita, **53**.
Passaloecus gracilis, **53**.
Passaloecus insignis, **53**, 54.
Passaloecus singularis, **53**.
Passaloecus turionum, **54**.
Pemphredon, 8, 10, 48, 50.
Pemphredon austriaca, **51**.
Pemphredon inornata, **51**.
Pemphredon lethifer, **51**.
Pemphredon lugubris, **51**.
Pemphredon morio, **51**.
Pemphredon rugifer, **52**.
Pemphredoninae, 48.
Pemphredonini, 8, 50.
Pepsinae, 23.
Pepsini, 23.
Philanthinae, 56.
Philanthini, 7, **58**.
Philanthus, 7.
Philanthus triangulum, **58**, 108.
Podalonia affinis, **36**.
Podalonia hirsuta, **36**.
Polistes dominulus, **33**.
Polistini, 33.
POMPILIDAE, 5, 13, **22**.
Pompilinae, 26.
Pompilus cinereus, **26**.
Ponera coarctata, **15**.
Ponerinae, 15.
Priocnemis agilis, **24**.
Priocnemis cordivalvata, **24**.
Priocnemis coriacea, **24**.
Priocnemis exaltata, **24**.
Priocnemis fennica, **25**.
Priocnemis gracilis, **25**.
Priocnemis hyalinata, **25**.
Priocnemis parvula, **25**.
Priocnemis perturbator, **25**.
Priocnemis pusilla, **26**.
Priocnemis schioedtei, **26**.
Priocnemis susterai, **26**.
Psenini, 48.
Psenulus, 48.
Psenulus concolor, **49**.
Psenulus pallipes, **49**, 53.
Psenulus schencki, **49**.
Pseudomalus auratus, **8**.
Pseudomalus violaceus, **8**.
Pseudospinolia neglecta, **12**, 30.
Psithyrus, see *Bombus*.

R
Rhopalum, 8.
Rhopalum clavipes, **47**.
Rhopalum coarctatum, **47**.

S
Sapyga quinquepunctata, 14, **15**, 89.
SAPYGIDAE, 13, **14**, 87.
Smicromyrme rufipes, 14.
Solenopsidini, 19.
Solenopsis fugax, 19.
SPHECIDAE, 2, **36**.
Sphecinae, **36**.
Sphecodes, 3, 63, 75, 76, 79, 87.
Sphecodes crassus, 79, **82**.
Sphecodes ephippius, **82**.
Sphecodes ferruginatus, **82**.
Sphecodes geoffrellus, 75, 79, **82**.
Sphecodes gibbus, **83**.
Sphecodes hyalinatus, 77, 78, **83**.
Sphecodes longulus, **83**.
Sphecodes miniatus, 79, **83**.
Sphecodes monilicornis, 76, 77, 79, **84**.
Sphecodes niger, 79, **84**.
Sphecodes pellucidus, 63, **84**.
Sphecodes puncticeps, **84**.
Sphecodes reticulatus, **84**.
Sphecodes rubicundus, 69, **85**.
Sphecodes scabricollis, **85**.
Sphecodes spinulosus, **85**.
[*Sphex,* 36].
Spilomena, 48.
Spilomena beata, **50**.
Spilomena curruca, **50**.
Spilomena enslini, **50**.
Spilomena troglodytes, **50**.
Spinolia neglecta, see *Pseudospinolia neglecta.*
Stelis, 3, 87.
Stelis breviuscula, **88**, 89.
Stelis ornatula, **88**.
Stelis phaeoptera, **88**.
Stelis punctulatissima, 87, **88**.
Stenamma westwoodii aggregate, 19.
Stenammini, 19.
Stigmus, 50.
Stigmus pendulus, **50**.
Stigmus solskyi, **51**.

[*Strongylognathus testaceus,* 4].
Symmorphus bifasciatus, **32**.
Symmorphus connexus, **33**.
Symmorphus crassicornis, 10, **33**.
Symmorphus gracilis, 30, **33**.

T
Tachysphex nitidus, **38**.
Tachysphex obscuripennis, **38**.
Tachysphex pompiliformis, 9, 11, **38**.
Tapinoma erraticum, **19**.
Temnothorax, 16.
Temnothorax albipennis, **18**.
Temnothorax interruptus, **18**.
Temnothorax nylanderi, **18**.
Tetramoriini, 18.
Tetramorium caespitum, 4, 16, **18**.
Tiphia, 13
Tiphia femorata, **13**.
Tiphia minuta, **13**.
TIPHIIDAE, 13.
Tiphiinae, 13.
Trichrysis cyanea, 9, **10**.
Trypoxylini, 37, **39**.
Trypoxylon, 8, 9, 10, 23.
Trypoxylon attenuatum, **39**.
Trypoxylon clavicerum, 10, **39**.
Trypoxylon figulus aggregate, **39**.
Trypoxylon figulus (sensu stricto), 39, **40**.
Trypoxylon medium, 39, **40**.
Trypoxylon minor, 39, **40**.

V
Vespa crabro, **34**.
VESPIDAE, 13, 30.
Vespinae, 33.
Vespini, 34.
VESPOIDEA, 2, **13**.
[*Vespula austriaca,* 4, 35].
Vespula germanica, 34, **35**.
Vespula rufa, 34, **35**.
Vespula vulgaris, 34, **35**.

X
Xylocopa violacea, **94**.
Xylocopini, 94.